Naturalists' Handbooks 18

Insects on cabbages and oilseed rape

WILLIAM D.J. KIRK
*Department of Biological Sciences,
Keele University*

With illustrations and plates by Miranda Gray

T0198070

Richmond Publishing Co. Ltd.

P.O. Box 963, Slough, SL2 3RS, England

Series editors
S. A. Corbet and R. H. L. Disney

Published by The Richmond Publishing Co. Ltd.,
P.O. Box 963, Slough, SL2 3RS
Telephone: Farnham Common (0753) 643104

Text © The Richmond Publishing Co. Ltd. 1992

Illustrations © Miranda Gray

ISBN 0 85546 287 6 Paper
ISBN 0 85546 288 4 Hardcovers

Printed in Great Britain

Contents

Editors' preface

Cabbages and related plants, known as brassicas, constitute a food supply that is unpalatable to most plant-feeding insects, because of poisonous chemicals, but which is exploited by some specialised insects that can tolerate these poisons. Among these specialists are the pests on cabbages and oilseed rape, and the natural enemies that help to keep down their numbers. Farmers and gardeners are familiar with the community of inects on brassicas, but until now few people have appreciated its fascination. Ubiquitous and accessible, it forms an excellent subject for the investigation of a range of ecological interactions, and it holds extra interest because of its potential importance for man. If we are to minimise the damage due to insect pests, those ecological interactions must be understood. If we interfere, for example by applying insecticides carelessly, without that understanding, we may fail to reduce pest damage, and may even tip the balance in a way that increases it.

This book introduces the insects on brassicas and their natural history, and shows how much remains to be discovered about even the commoner species. Until now, critical identification of brassica insects has been a task for experts, because it required specialised training and an extensive library. We hope that this book will help readers to name brassica insects more easily, and to explore for themselves this important and interesting community.

S.A.C.
R.H.L.D.
June 1991

Acknowledgements

I thank Keele University for supporting the writing and preparation of this book. I am grateful to Sally Corbet, Henry Disney, Maria Kirk and Chris Sutton for their comments on the text. The authorities of the Cambridge University Museum of Zoology, the Birmingham City Museum and Art Gallery, the Oxford University Museum and the City Museum and Art Gallery, Stoke-on-Trent allowed me to study their collections. H. Carter (Reading Museum and Art Gallery) helped me with the key to anthomyiid flies and A.H. Kirk-Spriggs (National Museum of Wales) helped me with the key to *Meligethes* beetles. I am also grateful for help and advice from: G. Burgess, J.M. Campbell, D. Emley, J.L. Evans, A.W. Ferguson, S.C. Littlewood, E.D. Macaulay, E.D. Morgan, M.G. Morris, A. Polwart, L.K. Ward, I.H. Williams, and A.L. Winfield. I thank the authorities of Harper Adams Agricultural College for allowing me to use their library.

W.D.J.K.

1 Introduction

A cabbage plant and an oilseed rape plant look very different from each other. One is short and fat and the other is tall and thin, so it may seem odd to consider them together. However, they are more alike than they look. Botanists classify them together in the same small group of species, the genus *Brassica*, in the family Cruciferae (also known as the Brassicaceae). Some of the best evidence that they are related comes from the similar form of the flowers and fruits (fig. 1), but since cabbages are normally eaten before they have a chance to flower and reproduce, their flowers are rarely seen. Other signs of relatedness can be found in their chemistry. For example, when the leaves are crushed, both plants give off the sharp sulphurous smell that is familiar from the cooking of cabbages. Chemical similarities are not immediately obvious to us, because our senses are dominated by sight, but insects depend much more on taste and smell. For them the plants are so similar that most of the insects that breed on one plant also breed on the other.

A rich community of insects lives on cabbages and oilseed rape, as gardeners and farmers know to their cost. Studies of these insects have revealed a fascinating world of interaction with their host plants and with each other. Many of these interactions are mediated by chemicals and so go unnoticed. This book describes some of these interactions and tells you how you can reveal them and investigate them further. The insects on cabbages and oilseed rape are particularly easy to study because the crops are common and there are usually plenty of insects on them. The plants are easy to grow in gardens. The insects that regularly occur on cabbages and oilseed rape can be identified with the keys in chapter 5.

There is also a practical reason why the interaction of insects with cabbages and oilseed rape should be of particular interest to us. We grow both these plants as crops, so we are closely involved with them. We cultivate, select and consume the plants and we try to kill or exploit the insects on them. We can therefore see ourselves as just one corner of an interacting triangle of three interested parties: insects, plants, and man. The insects are our rivals, competing for the food provided by the plants. Sometimes they win! A knowledge of the insect–plant interaction may help us tip the balance more in our favour.

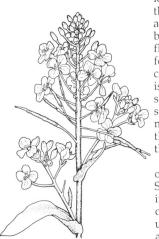

Fig. 1. A flowering stem of oilseed rape.

2 Brassicas as a place to live

The plants

Cabbages and oilseed rape look so different because
they have been artificially selected for different plant parts.
Cabbage has been selected for the ball of leaves in the bud at
the tip of the plant, while oilseed rape has been selected for the
seeds. The species in the genus *Brassica*, commonly referred to
as brassicas, and in the closely related genus *Sinapis* have
proved useful, mainly as sources of food for us and our
animals, and virtually every part of the plant has been
exploited (see table 1). The result is an array of crops with
exaggerated roots, stems, leaves, buds, flowers or seed pods
that would seem grotesque if we were not so used to them.
There has been much interbreeding to produce this diversity,
so the exact relationships between the crops are not always
clear. As a result, there is some debate about the correct
scientific names for each crop. A range of related crops and
wild plants are listed in table 1, together with their scientific
names. There are some surprises. Some crops that are very
different in appearance, such as cabbages, brussels sprouts
and cauliflowers, turn out to be varieties of the same species
selected for different parts, while some similar crops, such as
swede and turnip, are varieties of different species selected for
the same plant part. Radish and horseradish, with their hot-

Table 1. *Some examples of crops and wild plants in the genus* Brassica *and the closely related
genus* Sinapis *with the main parts of the plants for which the crops are grown. The format for
the scientific names of species is explained in chapter 5. The nomenclature follows the
classification used in the European Community (Anon., 1988, 1989) for crops and Clapham,
Tutin & Moore (1987) for wild plants. Note that* B. rapa *is also known as* B. campestris L.
Taxonomic categories below the level of species are abbreviated in the table:
ssp. = sub-species, convar. = convariety (a group of varieties), var. = variety

Species	Further classification	Common name	Part of plant used
Brassica oleracea L.	convar. *capitata* var. *alba*	Cabbage	Terminal leaf bud
	convar. *capitata* var. *rubra*	Red cabbage	Terminal leaf bud
	convar. *capitata* var. *sabauda*	Savoy cabbage	Terminal leaf bud
	convar. *oleracea* var. *gemmifera*	Brussels sprouts	Side leaf buds
	convar. *acephala* var. *sabellica*	Curly kale	Leaves
	convar. *acephala* var. *gongylodes*	Kohlrabi	Stem
	convar. *botrytis* var. *cymosa*	Sprouting broccoli, calabrese	Terminal flowers
	convar. *botrytis* var. *botrytis*	Cauliflower	Side flowers
	var. *oleracea*	Wild cabbage	—
B. napus L.	var. *oleifera*	Oilseed rape (swede rape)	Seeds
	var. *napobrassica*	Swede	Root
B. rapa L.	var. *sylvestris*	Turnip rape	Seeds
	var. *rapa*	Turnip	Root
B. juncea (L.) Czern.		Brown mustard	Seeds
B. nigra (L.) Koch		Black mustard	Seeds
Sinapis alba L.		White mustard	Seeds
S. arvensis L.		Wild mustard	—

flavoured roots, are also crucifers, but they are not in the genera *Brassica* or *Sinapis*. To identify crop species and their wild relatives, see Gill & Vear (1980) and Clapham, Tutin & Moore (1987)*.

The long history of the cabbage crop is described in Simmonds (1976); cultivated forms of cabbage were grown at least 2,000 years ago in the Mediterranean region. The leaves are usually cooked fresh, pickled as in sauerkraut, or used uncooked in salads. Although the crop separated from its wild ancestors long ago, we can still have some idea of what these ancestors were like by studying wild cabbage (*B. oleracea* var. *oleracea*), which grows on sea cliffs in southern England and Wales.

Oilseed rape is a general name that has been used for oil-seed crops of several species of *Brassica*. In Britain, the name is usually taken to refer to the dominant oil-seed crop, swede rape (*B. napus*), and this is the sense in which the name is used in this book. In parts of Canada and Scandinavia, a significant proportion of the oil-seed crop is turnip rape (*B. rapa*), which looks very similar. Oilseed rape is commonly referred to simply as rape, the word coming from the Latin *rapum*, which seems to have been the name for turnips and turnip rape.

The history of oilseed rape is confused because in the past it has not been distinguished from turnip rape. However, the crop is known to have spread within Europe and reached Britain in the 16th century. Evidence from the chromosomes shows that the plant is the result of hybridisation between *B. oleracea* and *B. rapa*. There do not seem to be any wild ancestral forms. The plant was grown mainly for the seed, which was crushed to produce oil which could be burned in lamps. The leftover seed meal was fed to animals. By the end of the 19th century, the oil was no longer needed for lighting, and demand was restricted to a few industrial uses, such as lubrication. Most of the crop was grown for animal forage rather than for seed. However, during the 1970s there was a phenomenal increase in the area of rape grown for oil-seed in Britain. Several factors were involved: the world price of rape seed rose substantially; the Common Agricultural Policy of the European Economic Community guaranteed the price; oilseed rape could be used as a break crop for cereals, interrupting the build-up of pests and diseases, while using much of the same farm machinery, such as combine harvesters; and there was increased demand for poly-unsaturated vegetable fats to reduce the risk of heart disease. The history of the crop, including these developments, is described by Ward and others (1985) and Scarisbrick & Daniels (1986).

In Britain, most of the crop is winter rape, which is sown in August or September, flowers from May to June, and is harvested in July or August. Some spring rape is also grown. This can be sown from March to early May. It flowers from June to July and is harvested in September. Rape is

*References cited under the authors' names in the text appear in full in the reference list on p.61.

harvested in an unusual way; the crop is often cut (swathed) or sprayed with a desiccant and allowed to dry in the field before harvesting by combine harvester. Cultivation and harvesting practices are described by Ward and others (1985).

Today, the oil is used for margarine, cooking oils and fats, and several industrial processes, and the meal is included in the feed of cattle, pigs, and poultry. Unfortunately, some of the natural chemicals in the oil and meal have undesirable effects. First, erucic acid in the oil has been found to produce heart lesions in laboratory animals. Plant breeders have tackled the problem by producing "single low" varieties with low levels of erucic acid in the seeds and these started to be widely grown in the 1970s. Secondly, the glucosinolates in the seed meal break down to give unpalatable, toxic and thyroid-damaging (goitrogenic) substances, so that meal containing high concentrations can be incorporated into animal feeds in only small amounts. Plant breeders were able to produce "double low" varieties with low levels of both glucosinolates and erucic acid in the seeds, and these began to be widely grown in Europe by the end of the 1980s. There is no shortage of suggestions for future changes to the content of the seeds. "Triple low" varieties, already grown in Canada, have a low fibre content in the seed coat to improve palatability to animals. We may yet see a decrease in linolenic acid, which can give an unpleasant smell to the oil, or an increase in linoleic acid (vitamin F) because of its nutritional value.

Of these substances, those most characteristic of the family Cruciferae are the glucosinolates. Not only do they matter to us, but they are of major significance to the insects on cabbages and oilseed rape.

The glucosinolates

These sulphur-containing compounds are important because of their breakdown products. The tissues of crucifer plants contain glucosinolates and, stored separately, an enzyme called myrosinase. When the tissues are crushed or chewed, the enzyme comes into contact with the glucosinolates and a variety of volatile compounds are released. These are mainly isothiocyanates, also known as mustard oils. They are released from the intact plant as well, but at much lower concentrations. Usually, the same glucosinolates occur throughout the plant, but concentrations vary between different parts of the plant and change as the plant grows.

The best-known glucosinolate is sinigrin, which occurs at particularly high concentrations in the seed of black mustard (*Brassica nigra*). It takes its name from the first parts of the old scientific name for this species, which was *Sinapis nigra*. The volatile breakdown product is 2-propenyl isothiocyanate. (Many publications use the alternative name for this substance – allyl isothiocyanate.) This volatile gives the slightly bitter flavour of cabbages and brussels sprouts at low concentration and the hot flavour of mustard at higher concentration. It is these volatile isothiocyanates that make the

Fig. 2. The chemical reaction when sinigrin, a glucosinolate, breaks down in the crushed tissue of crucifer plants. The 2-propenyl isothiocyanate is volatile and has a strong flavour. Glucosinolate and isothiocyanate end-groups are enclosed in boxes for clarity. C is carbon, H is hydrogen, O is oxygen, S is sulphur, and N is nitrogen. The dashes show the number of bonds between the main atoms.

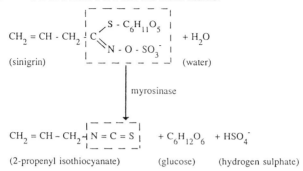

brassicas taste interesting or, for some people, unpleasant! The reaction that takes place in your mouth is shown in fig. 2.

The diagram of sinigrin in fig. 2 shows the typical structure of a glucosinolate. There is the characteristic glucosinolate end-group, highlighted by a box, and then a side-chain. The action of the enzyme changes this complicated end-group to the simpler isothiocyanate ending, while leaving the side-chain unchanged. All glucosinolates have the same end-group but they vary in the shape and length of the side-chain. Those with side-chains similar to that of sinigrin (unbranched and with a double bond between two carbon atoms) are called the alkenyl glucosinolates, and they are commonly found in the genus *Brassica*. Table 2 lists the major

Table 2. *The major alkenyl glucosinolates and their volatile breakdown products in cabbages and oilseed rape (chemists will note that in glucosinolates with an -OH group in the second position of their side-chain, the initial product is an isothiocyanate, but the -OH reacts with the -NCS to give an oxazolidine-2-thione)*

Glucosinolate	Glucosinolate side-chain	Main breakdown product
Cabbage:		
Sinigrin	$CH_2= CH-CH_2-$	2-propenyl isothiocyanate
Gluconapin	$CH_2=CH-CH_2-CH_2-$	3-butenyl isothiocyanate
Oilseed rape:		
Gluconapin	$CH_2=CH-CH_2-CH_2-$	3-butenyl isothiocyanate
Glucobrassicanapin	$CH_2=CH-CH_2-CH_2-CH_2-$	4-pentenyl isothiocyanate
Progoitrin	$CH_2=CH-CHOH-CH_2-$	5-vinyloxazolidine-2-thione
Gluconapoleiferin	$CH_2=CH-CH_2-CHOH-CH_2-$	5-allyloxazolidine-2-thione

alkenyl glucosinolates in cabbages and oilseed rape, and their main breakdown products. It shows that the two plants differ in some of their alkenyl glucosinolates, but have some in common, and that they both give off similar isothiocyanates. Many experiments with insects on cabbages and oilseed rape have used sinigrin, because this glucosinolate is available commercially (see p. 59). Although sinigrin is not present in significant amounts in oilseed rape, very similar compounds are, and insects are likely to respond to them and their volatiles in similar ways.

Cabbages and oilseed rape also contain several other types of glucosinolate, many of which also break down to give volatile isothiocyanates. Further details can be found in Fenwick and others (1983). For those interested in the chemistry of glucosinolates, their breakdown products, and other crucifer volatiles, Vaughan and others (1976) provide a good review.

Although the toxic breakdown products are a problem in animal feed, they are not a problem for us because crucifers make up only a small part of our diet and the plant parts that we eat in any quantity have relatively low concentrations of glucosinolates, perhaps because of selection against bitterness at an early stage in domestication.

The release of unpalatable and toxic substances when the tissue is damaged suggests that glucosinolates are defensive compounds. The substances are unpalatable and toxic not only to vertebrates, but also to insects and slugs, and they inhibit bacterial and fungal diseases.

The insects

One might doubt that glucosinolates are unpalatable or toxic to insects, since many insects thrive on cabbages and oilseed rape! But most of these insects are brassica specialists and presumably have special adaptations that enable them to overcome any chemical defences. To see the deterrent or toxic effects of glucosinolates, we should look at insects that feed on other plants, rather than at brassica specialists. To do this, we cannot just transfer an insect onto a cabbage leaf and see whether the insect starves or is poisoned. Many insects will not feed unless stimulated to do so by certain chemical characteristics of their usual host plant. If an insect refuses to feed on cabbage, it may be because of the absence of a stimulant host plant chemical, rather than the presence of deterrent glucosinolates. Erickson & Feeny (1974) carried out an elegant experiment that avoided this problem. Caterpillars of the black swallowtail butterfly (*Papilio polyxenes* (F.)) feed on the leaves of celery and related plants in the family Umbelliferae, which do not contain glucosinolates. Celery leaves were made to take up sinigrin by putting the cut stems in solutions of sinigrin in water. The treated leaves were then fed to the caterpillars. When the concentration was similar to the natural level of glucosinolates in brassicas (about 0.1% by weight), all the caterpillars died before pupation.

The breakdown product, 2-propenyl isothiocyanate, was detected in the gut contents and body tissues of the caterpillars; it is likely to have been responsible for the deaths. Caterpillars on untreated leaves survived. The result showed that sinigrin can kill a non-brassica insect, but it would be useful to test a wider range of insects. What happens with beetles, fly larvae or aphids? Further experiments by Blau and others (1978) showed that sinigrin at natural concentrations made foliage less palatable (less was eaten) and also toxic (caterpillars showed less growth for the same consumption) to the black swallowtail, but sinigrin had no effect on the small white butterfly (*Artogeia rapae*), a brassica specialist, even at unusually high concentrations. However, it was mildly toxic to the southern armyworm moth (*Spodoptera eridania* (Cramer)), a generalist, which feeds on a wide range of plants.

It is generally assumed that glucosinolates are toxic to most non-brassica insects, which means it would be an evolutionarily difficult step for an insect to start feeding on brassicas. The insects on brassicas have presumably taken this step by evolving a way of detoxifying the glucosinolates. We might ask why most of these insects are restricted to brassicas. Why not expand the host range by adding further detoxification systems? Perhaps each biochemical detoxification system is costly in terms of energy and the insects can afford to have only one or two. Or perhaps increased competition with other insects would outweigh the potential advantage of a wider host range.

Some specialist insects have gone a step further in the evolutionary battle between brassicas and the insects that damage them. They not only overcome the plants' chemical defences, but, as we shall see in chapter 3, they use them to stimulate egg-laying (oviposition) and feeding, and they use the volatiles to locate the plants from a distance. How might the plants evolve to fight back? How will the use of varieties low in defensive glucosinolates affect the insect–brassica battle?

Many brassica insects have specialised further and are restricted to a particular plant part, and this is the same whatever crop they are on. For example, there is a cabbage *root* fly, a cabbage *leaf* miner, and a cabbage *seed* weevil. In chapters 3 and 4, we shall see examples of specialised behaviours for locating particular parts of the host plant.

3 The herbivores

The large white butterfly (*Pieris brassicae*)

Fig. 3. A cluster of eggs laid by a large white butterfly.

This insect, commonly known as the cabbage white, is probably the best-known cabbage pest. It should not be confused with the small white (*Artogeia* (= *Pieris*) *rapae*), the other common cabbage butterfly. An adult female is very conspicuous when she is fluttering around over cabbages. She lands on a leaf and drums on it with her feet. Sometimes she curls her abdomen under the edge of a leaf and lays a batch of about 30–100 eggs (fig. 3). The eggs are yellow, just over 1 mm tall, and skittle-shaped with about 17 vertical ridges.

The adult drums on the leaves to test for the presence of glucosinolates by means of sensory cells on her feet. If a glucosinolate stimulus is produced artificially by brushing her feet with sinigrin solution, she can be induced to lay eggs in the absence of brassicas. Other substances may also be involved in host recognition, but the presence of glucosinolates is the common factor linking the main host plants. One of the most commonly used hosts, after cabbage, is the garden nasturtium (*Tropaeolum majus* L.), in the South American family Tropaeolaceae. Cabbages and nasturtiums are not closely related, but they both contain glucosinolates.

An egg-laying female can "assess" the egg load on a leaf and avoid overloading it, reducing the risk of food shortage for her offspring. Rothschild & Schoonhoven (1977) found that adults fluttering above the plant were deterred from landing by the sight of eggs, as well as by a chemical marker left on the eggs by the female that had laid them. Fewer eggs were laid on leaves with yellow plastic model eggs or crushed-egg fluid. There is much interest in such messenger chemicals (semiochemicals) because they can perhaps be used in conjunction with, or even instead of, insecticides to reduce the impact on the environment (Pickett, 1989). Semiochemicals are known as pheromones when they carry a message between members of the same species, as this egg marker does, and as allelochemics when they carry a message from one species to another, as sinigrin does for large white butterflies. When grown near cabbages, some plants, such as tomato (*Lycopersicon esculentum* Mill.) and thyme (*Thymus vulgaris* L.), are thought to deter large whites from laying eggs on the cabbages (review in Feltwell, 1982); perhaps their scent is acting as a deterrent. The effects of plant associations and the use of companion plants are discussed on p. 16.

The larvae (pl. 2.1) are stimulated to feed when sensory cells on the mouthparts come in contact with glucosinolates. The larvae are gregarious at first, but when they have moulted four times and are fully grown, they enter a solitary "wandering stage", during which they can walk long distances in search of a pupation site. They select vertical surfaces away from the host plant, such as tree

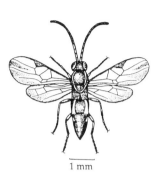

1 mm

Fig. 4. An adult parasitoid wasp, *Apanteles glomeratus*.

Fig. 5. A large white butterfly caterpillar parasitised by *Apanteles glomeratus*.

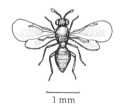

1 mm

Fig. 6. An adult hyperparasitoid, *Tetrastichus galactopus*.

trunks, fences and walls. There are one or two generations per year in Britain and the insect overwinters as a pupa.

Many caterpillars are parasitised by a small black parasitic wasp, *Apanteles glomeratus*, in the family Braconidae (fig. 4). An adult female wasp will pierce young caterpillars with her ovipositor (egg-laying tube) and inject many eggs. The wasp larvae hatch and feed on the caterpillar from within. When the caterpillar is fully grown, the wasp larvae pierce small holes in its skin and emerge. The emergence is synchronised and takes only a few minutes. The larvae then spin yellowish white cocoons in which to pupate, forming clumps of about 30 around the dead or dying caterpillar (fig. 5). The adult wasps may appear within a few weeks or wait until the following spring. Insects like *A. glomeratus* that develop at the expense of a single individual host animal, usually killing it, and have a free-living adult, are known as parasitoids. They are easily overlooked, since they are usually small, but they are important because they are natural biological-control agents that decrease the size of pest populations. They can be caught in yellow water-traps or reared out of parasitised insects (techniques p. 58).

Even more easily overlooked are the hyper-parasitoids. These are parasitoids of parasitoids, and they are even smaller! *Tetrastichus galactopus* is a minute metallic green parasitic wasp in the family Eulophidae (fig. 6). It lays its eggs in larvae of *A. glomeratus* that either are still within the caterpillar or have emerged but not yet pupated. *Tetrastichus* adults eventually emerge from the *A. glomeratus* cocoons. Another common parasitic wasp that can emerge from the cocoons is *Lysibia nana* in the family Ichneumonidae. It lays its eggs in *A. glomeratus*, but in the cocoons instead of in the larvae. Littlewood (1988) collected 4,822 cocoons of *A. glomeratus* from gardens in Shropshire and reared out 938 *A. glomeratus*, 949 *T. galactopus*, 1,917 *L. nana*, and 177 individuals of other species. It would be interesting to see how the percentage parasitism varies with plot size or between plant hosts. Richards (1940) found that a lower percentage of caterpillars were parasitised on nasturtium (*Tropaeolum majus*) than on cabbages. This could be investigated further.

Large white caterpillars can be a serious pest, despite high mortality from parasitoids, and various chemical insecticides are used against them. Many strains of the naturally occurring bacterium *Bacillus thuringiensis* Berliner, known as Bt, produce substances which are toxic when eaten by butterfly and moth caterpillars, but are not harmful to other insects or to mammals. Bt can therefore be cultured and sprayed on crops as a selective biological-control agent. A Bt spray may be acceptable to organic growers because it is of natural origin, selective, and without harmful residues. However, on small plots, the cheapest and easiest way to prevent damage is often to pick off caterpillars by hand!

The cabbage root fly (*Delia radicum*)

The adults of this abundant pest are greyish, slightly hairy flies about 6 mm long; they look much like houseflies (pl. 3.4). In southern England, they emerge in late April and May from pupae that have overwintered in the soil. The females feed on the nectar of plants such as cow parsley (*Anthriscus sylvestris* (L.) Hoffm.) and disperse to crucifer crops, where they lay batches of eggs among soil particles at the base of the stems. The eggs are white and 1 mm long. The yellowish white larvae (pl. 5.6) burrow into the roots and feed on them for three to four weeks. When they are about 10 mm long, they pupate in the soil nearby. They are most damaging to young plants. Damage is evident when the older leaves develop a reddish or purplish tinge and wilt. A second generation of adults emerges in July and there may also be a small third generation in southern England in late August and early September. The biology is reviewed by Coaker & Finch (1971).

A female locates host plants by flying upwind in a series of short flights when exposed to host plant volatiles. This behaviour eventually brings her to the source of the scent. The effectiveness of different volatiles can be measured in the field by seeing how much they increase the catches of water traps (techniques p. 58). Finch & Skinner (1982) found a range of isothiocyanates that significantly increased trap catches, including 2-propenyl isothiocyanate, the breakdown product of sinigrin, but these isothiocyanates were effective only in unnaturally high quantities measured in grams. Concentrated extracts of swede roots were effective in milligram quantities. This finding supports the idea that the flies respond best to a blend or "bouquet" of volatiles, not to just one. Cabbage root flies land in response to wavelengths of light concentrated in the region of yellow and green (500–600 nm), as reflected by green foliage. This response normally enables them to distinguish between foliage and other visual features (the sky and the soil), but it means that relatively few flies land on red cabbage. These visual and olfactory responses can be exploited to increase trap catches by painting traps yellow, or better still fluorescent yellow, and adding host plant extract or 2-propenyl isothiocyanate. Direct control of cabbage root flies by mass trapping with such traps has had little success, but it might work on very small plots. Large numbers of yellow washing-up bowls, filled with water and a few drops of detergent, could be arranged between the plants. Unfortunately, these traps will also catch beneficial predators and parasitoid wasps.

A female fly lands on a leaf of a young plant when she is ready to lay eggs, and goes through a sequence of behaviour before eventually laying (fig. 7) (Zohren, 1968). First, she walks along the leaf, pausing now and then to groom or make short flights. This is the latent phase. Then she walks continuously, often around the edge of the leaf, frequently changing direction (leaf-blade walking). She

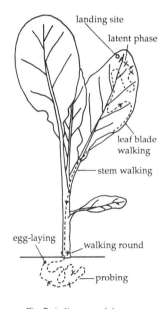

landing site
latent phase
leaf blade walking
stem walking
egg-laying
walking round
probing

Fig. 7. A diagram of the egg-laying behaviour of adult female cabbage root flies on young brassica plants (after Zohren, 1968).

appears to be using sensory cells in her feet to assess the suitability of the plant. The cells in the feet can detect host extracts, sinigrin, and other substances. The fly will not lay eggs without such stimuli. If she moves onto a stem or the midrib of a leaf, she follows it quickly to the ground (stem walking). When she reaches the base of the stem, she moves round it sideways, keeping her head towards the ground (walking round). Then she moves onto the soil and walks around close to the stem, occasionally climbing up the stem a few centimetres (climbing). The fly starts to bend her abdomen down and probe the soil at intervals with her ovipositor, perhaps testing soil suitability. Finally, she lays her eggs in the soil close to the stem, jerking her body and scraping soil particles around the ovipositor with her hindlegs (fig. 8). Most eggs are laid in dry soil with particles about 1 mm across.

Fig. 8. An adult female cabbage root fly laying eggs at the base of a brassica stem.

A fly may not always perform this complete sequence. She may break off and fly away at any time. Sometimes during climbing she walks back up to the leaves and begins leaf-blade walking again. If the behaviour of many flies is followed, the frequency of successful transitions between each behavioural stage and the next can be scored. These transition frequencies could be compared between plants with different levels of infestation. Comparisons could be of the same plant variety at different sites or of different plant varieties at the same site. A large discrepancy in one particular transition frequency might suggest a cause of the difference in infestation. For example, a low frequency of transitions from probing to egg-laying would suggest poor soil suitability. The results could indicate plant factors that give some species or varieties more resistance to cabbage root fly than others.

The presence of other insects can deter egg-laying. Aphids seem to do this by physically interrupting the flies on the leaves (Finch & Jones, 1989), while caterpillars of the garden pebble moth (*Evergestis forficalis*) (pl. 3.2) have a deterrent chemical (sinapic acid) in their frass (Jones and others, 1988).

In a small plot, it is easy to reduce the number of eggs laid on cabbages by placing around the plant stem 12 cm diameter discs of felt or carpet underlay, with a slit and a 0.5 cm hole in the middle (fig. 9) (Skinner & Finch, 1986). These make it difficult for the flies to touch the soil particles and receive the right stimuli for egg-laying. The discs also act as a mulch and conserve moisture for the plant, enhancing growth. Predatory ground beetles (Carabidae) and rove beetles (Staphylinidae) aggregate in the moist, sheltered area under the discs and further reduce herbivore numbers by predation. In large-scale crops, the disc method is not usually economical. Instead, granular insecticides are often applied to the soil.

Fig. 9. A felt disc used to reduce cabbage root fly infestation.

The cabbage seed weevil (*Ceutorhynchus assimilis*)

The adult weevils are grey, about 3 mm long, with a distinct snout (rostrum) (pl. 1.6). After overwintering in

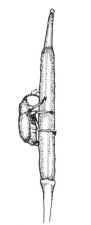

Fig. 10. Brushing behaviour of a female cabbage seed weevil on a pod.

Fig. 11. A pod with the exit hole of a cabbage seed weevil.

Fig. 12. An open pod with pod midge larvae inside.

debris or soil, they fly to flowering crucifers from late April to June. Host plant odours seem to be involved in host finding, since they increase trap catches (Free & Williams, 1978). By inserting delicate electrodes into sensory cells in the antennae, Blight and others (1989) found cells that gave strong responses to 3-butenyl isothiocyanate and 4-butenyl isothiocyanate and weak responses to 2-propenyl isothiocyanate. These volatiles are given off by cabbages and oilseed rape (table 2). When adults begin flying into fields of oilseed rape, they are commonest at the edges, so if spraying is necessary at this time it could be restricted to crop borders. Problems of spraying a flowering crop are discussed in the next chapter.

A female weevil feeds on pollen until her ovaries have developed and then she moves to developing pods about 20–40 mm long. Here she bores a hole through the pod wall with her snout and then eats the contents of seeds within the pod. The typical egg-laying behaviour exploits this ability to bore holes (Kozlowski and others, 1983). A female walks along the stem to a pod and then walks slowly along it to the far end and back while tapping the pod surface with her antennae. About 1–5 mm from the stem, she makes a hole in the pod wall with her snout, as for feeding. She turns round so that the ovipositor is above the hole and inserts her ovipositor for about 2 minutes to lay a single egg. When she has withdrawn her ovipositor, part of the ovipositor is exposed and drawn along the pod surface in zigzags as the female walks in a spiral around the pod to the end and back (fig. 10). She then walks away along the stem. Each female can lay many eggs in this way. This brushing behaviour leaves behind a marker pheromone which discourages other females from laying eggs there. This has a clear advantage in wild hosts with small pods providing enough food for only one larva, but it may not matter so much in the very large pods of oilseed rape, each of which could support several larvae.

The white larva (pl. 5.2) reaches its full length of about 5 mm after feeding on about five seeds within the pod. Four weeks after the egg was laid, the larva bores through the pod wall (fig. 11) and drops to the ground, where it pupates in a cocoon a few centimetres below the soil surface. The next generation of adults emerges two to three weeks later and feeds before overwintering.

The damage from seed destruction by the cabbage seed weevil is complicated by an interaction with the brassica pod midge (Dasineura brassicae). The midge (pl. 4.5), which is only 1–2 mm long, has a weak ovipositor and can only lay its eggs in pods that have already been damaged, so it makes use of the feeding and oviposition holes of the seed weevil. It lays batches of about 20 eggs in each pod. The whitish cream larvae (pl. 5.5) are up to 2 mm long and feed on the inside surface of the pod wall (fig. 12), causing "bladder pod" symptoms. The pods swell slightly, turn yellow, and ripen and split prematurely, spilling all the seeds before harvest. The larvae drop to the ground to pupate. There are several generations per year and they overwinter as larvae.

If entomologists could predict the damage to be caused by seed weevil, they could help farmers make decisions about pest management, such as whether it is economic to spray the crop. It would be useful if yield loss could be predicted from the number of insects per plant, but there are several problems. Oilseed rape plants have an enormous capacity to compensate for damage by modifying their growth, so there is not a simple relationship between insect numbers and yield, or even between the number of damaged pods and yield. Plants can respond to damage by producing more flowers or increasing the weight of the remaining seeds, so that large numbers of flowers or pods can be removed without affecting the final yield of the plant (Williams & Free, 1979). However, compensatory growth extends the period of flowering, and so produces a crop that does not ripen uniformly. This can result in large losses at harvest; immature pods will give no ripe seed and over-ripe pods will shatter and spill their seed to the ground as they are combine harvested. There can be an additional problem if the spilt seed produces "volunteer" rape plants amongst the following crop next year. Pod midge causes serious spillage of seed, so seed weevil becomes much more of a problem if pod midge is present.

In order to decide whether an insecticide is worthwhile for a particular level of insect pest damage, entomologists must carry out experiments to find the infestation level at which the benefit of insecticide treatment (the value of the predicted saving in yield) exceeds the treatment costs (including insecticide and yield losses through damage from spraying machinery). A level of one or more weevils per plant has been suggested; this threshold may need to be lower if pod midge is present (Free and others, 1983).

The community

The community of insects associated with cabbages and oilseed rape has a complicated network of interactions. Ecologists simplify the study of communities by grouping together species that exploit similar resources in similar ways. Guilds are groups of species united by function rather than relatedness. Species belonging to the same guild may interact with each other particularly strongly because they have similar modes of life, whereas members of different guilds are expected to have weaker interactions. This system was first used by Root (1973), who was studying communities on leafy cabbages, known as collards, in the USA. He divided the insect herbivores into "strip feeders" that chew leaves (for example butterfly caterpillars), "pit feeders" that rasp small pits or holes in the leaves (for example adult flea beetles), and "sap feeders" that suck the sap (for example aphids).

Table 3. *A hypothetical niche diagram for phytophagous insects feeding on a species of brassica at one site. The rows and columns give the possible feeding sites and feeding methods. Each dot represents one species of insect. Species that feed on more than one plant are represented by dots connected by lines.*

	Chew	Suck	Mine	Gall
Root		• • •		•
Stem		•	• •	
Leaf	• • • • •	•	•	
Flower	•			
Pod	•	•		

Lawton (1982) divided up herbivorous insect species on bracken (*Pteridium aquilinum* (L.) Kuhn) according to whether they chewed, sucked, mined or galled the plant, and then according to which of several parts of the plant they ate. A grid represented all combinations of feeding method and plant part. Table 3 shows a hypothetical example of a similar grid. Each box represented a different way of feeding, and thus a different potential ecological niche. A niche is a set of resources and environmental conditions that can be used by a species. Much ecological theory has been built around this concept (Begon and others, 1990). For example, it has been suggested that the number of potential niches on a plant species can affect the number of herbivore species that live on it. Although insect species may differ from one another in many different aspects of habitat usage, for simplicity it is common to limit consideration to the food resource. Lawton put a dot in the appropriate box or boxes for each insect species that was present. The completed niche diagram revealed vacant niches (boxes with no dots) and niches where there may have been inter-specific competition for food (boxes with many dots). This is a useful way of representing the structure of a community and comparing communities on different plant species or on the same species at different sites. The work is described and discussed by Strong and others (1984) and Begon and others (1990).

Similar niche diagrams could be produced for insects that chew, suck, mine or gall the roots, stems, leaves, flowers or pods of brassicas (table 3). The diagrams could be used to explore differences between cabbages and oilseed rape, between gardens and farms, between open and sheltered sites or between years.

Inter-specific competition occurs when one species harms another by depriving it of a resource. For this to be of evolutionary significance, the harm must reduce the number or quality of the victim's offspring. If two species are in the same box in a niche diagram, they share the same resource in the same way, so they are likely to compete. But this is not necessarily so; there may be enough food for both. To demonstrate competition it is necessary to show that a species does better when its competitor is removed or that it does worse when the numbers of its competitor are supplemented. Manipulative experiments of this kind have been done with herbivores on cabbages. When Kareiva (1982) removed eggs and caterpillars of the small white butterfly (*Artogeia rapae*) from densely planted collards in the USA, the numbers of an adult flea beetle (*Phyllotreta cruciferae*) increased. When this flea beetle was removed, the numbers of another flea beetle (*Phyllotreta striolata* (F.)) increased. There is plenty of scope for further experiments of this kind to investigate competitive relationships among the insect species on cabbages and oilseed rape.

Another reason why two species in the same box may not be competing is that their resource may be further sub-divided. For example, caterpillars of the large white and the small white both chew cabbage leaves, but the large white feeds mainly on the outer leaves, while the small white feeds mainly on the inner heart leaves. This dietary separation may reduce competition, but may not eliminate it altogether. It has been suggested that all the herbivores on a plant can be considered as competing ultimately for the same finite supply of resources which can be transferred between plant parts. For example, leaf feeders may take nutrients that would otherwise have gone to the roots. This interesting idea needs further testing. For example, is the growth and survival of cabbage root fly larvae affected by the presence of large white caterpillars on the leaves? If so, at what level of infestation does the effect become apparent?

Insect herbivores can damage a crop and reduce the yield. How can we manage a crop to reduce the damage? Any action we take against the main pest species is likely to affect the whole insect community, including the interactions between the plant and the herbivores, parasitoids, predators, and hyperparasitoids. For example, a general insecticide may kill parasitoids and predators as well as herbivores. It will give an immediate reduction in pest herbivore numbers, but this could be followed by a pest outbreak if the remaining herbivores then multiply unhindered by parasitism and predation. To preserve interactions within the community that reduce the risk of subsequent outbreaks, we try to use methods of pest management that avoid killing non-target organisms. Specific insecticides are therefore better than non-specific ones.

It may also be possible to reduce pest damage by adapting methods of cultivation. One way to do it is to grow different crops together, usually in alternate rows. This intercropping is not usually practicable in large-scale agriculture, but it is common in small plots and gardens, especially in the tropics. One advantage is that the two species of crop will differ slightly in their requirements, and the resulting reduction in competition can give higher yields. Another is that pest herbivore numbers are often lower. For example, Root (1973) found that collards grown among many other plant species had fewer herbivores on them than collards in a pure stand, and Ryan and others (1980) found fewer cabbage root fly larvae in cabbage intercropped with clover or lettuce than in cabbage grown without other plants. Intercrops do not have to be vegetables. French marigolds (*Tagetes patula* L.) have been suggested as companion plants to reduce pest damage on cabbages (Philbrick & Gregg, 1967).

There has been much debate over the interpretation of such results (Price, 1984). Why do mixed plantings suffer less pest damage? Is it because they retain more predators and parasitoids? Or is it because they make it harder for the egg-laying female pests to find their host plants? Does a flowering plant grown between cabbages draw nectar-feeding adult parasitoids into the crop and so increase herbivore mortality? It would be useful to know which combinations of plant species are best at reducing numbers of each of the pest species, and how this works. Care must be taken to distinguish the effects on yield of changes in planting pattern, which would happen in the absence of insects, from effects of changes in insect damage.

Even plot size can affect insect density. Cromartie (1975) grew plots of collards in the USA with 1, 10 and 100 plants and counted the numbers of insects per plant. The small white butterfly was commonest on single plants, adults of the flea beetle *Phyllotreta striolata* were commonest on 10-plant patches, and another flea beetle, *P. cruciferae*, was commonest on 100-plant patches. Such plot-size effects will produce differences in infestation between gardens and farms. The result for the small white can be explained in terms of movement patterns of egg-laying adult females (Jones, 1977), but other mechanisms are likely to be involved for other species.

Table 4 shows how the community changes with season. Most pests are common on both cabbages and rape, but six species that live on leaves are much commoner on cabbages. The three pests of flowers and pods are not pests on cabbages because the plants are harvested before flowering, but they will breed on cabbages that are allowed to flower. Time of planting, as well as type of crop, can affect pest numbers. For example, rape winter stem weevil and cabbage stem flea beetle lay eggs on plants in autumn and winter, so they affect winter rape but not spring rape. Adult flea beetles particularly attack young seedlings in spring, so they affect spring rape but not the older winter rape.

Table 4. *The months when the main pests of cabbages and oilseed rape are commonest on the crops in Britain. The pests are divided according to whether they are mainly pests of cabbages, oilseed rape or both. The pests may be present in smaller numbers at other times.*
Key: a = *adults,* l = *larvae*

Insect pest	Month											
	J	F	M	A	M	J	J	A	S	O	N	D
Cabbages:												
Large white (*Pieris brassicae*)					a	a	al	al	al	al	l	
Small white (*Artogeia rapae*)				a	a	l	al	al	al			
Cabbage moth (*Mamestra brassicae*)					a	al	al	al	al	al		
Garden pebble moth (*Evergestis forficalis*)					a	al	al	al	al		l	
Diamond-back moth (*Plutella xylostella*)					al	al	al	l	l	l		
Cabbage whitefly (*Aleyrodes proletella*)	a	a	a	a	al	al	al	al	al	al	a	a
Oilseed rape:												
Pod midge (*Dasineura brassicae*)					al	al	al	al	al		l	
Pollen beetles (*Meligethes* species)					a	al	al	al	a			
Cabbage seed weevil (*Ceutorhynchus assimilis*)					a	al	al	al	al			
Rape winter stem weevil (*Ceutorhynchus picitarsis*)	al	al	al	l	l				a	al	al	al
Cabbages and oilseed rape:												
Cabbage root fly (*Delia radicum*)					al	al	al	al	al	al	l	
Cabbage leaf miner (*Phytomyza rufipes*)					a	al	al	al	al	al		l
Cabbage aphid (*Brevicoryne brassicae*)					al	al	al	al	al	al	al	
Peach-potato aphid (*Myzus persicae*)				al	al	al	al	al	al	al	al	
Cabbage stem flea beetle (*Psylliodes chrysocephala*)	l	l	l	l	l	a	a	a	a	al	al	l
Cabbage stem weevil (*Ceutorhynchus quadridens*)					a	al	al					
Flea beetles (*Phyllotreta* species)					a	al	al	al	al			

4 The flower and its visitors

The flower

The flowers of all the species in the family Cruciferae have four petals arranged in the form of a cross. This characteristic is recorded in the name Cruciferae, which means "cross bearer". The flowers of cabbages and oilseed rape are alike (fig. 13). The yellow petals are surrounded by four green sepals. Inside the flower are six stamens, which produce pollen. In the centre of the flower are the female parts: the ovary which will develop into a pod (the botanical term siliqua is often used for a crucifer pod), and above it the long style ending in the stigma. A pollen grain that lands on the stigma germinates, producing a pollen tube which grows down to the ovary to fertilise the ovules, which can then develop into seeds.

Fig. 13. A fully open oilseed rape flower.

The flower has stamens of two different types. The four inner stamens are long, reaching beyond the level of the petals and releasing their pollen outwards. The two outer stamens are short, releasing their pollen inwards below the level of the petals. There are four nectaries just outside the gaps between the long stamens. These nectaries are small dark green bumps, often surrounded by a clear drop of the nectar they have secreted. The two behind the short stamens are the inner nectaries, and the exposed ones are the outer nectaries. The nectar secreted by the inner nectaries is more copious and often less concentrated than that of the outer nectaries, and some insects are choosy about which type they visit. It is sometimes possible to see whether an insect is visiting the inner nectaries, the outer nectaries, or both (Williams, 1980).

In many species of plant, outbreeding is promoted by a biochemical recognition system that inhibits self-fertilisation by a plant's own pollen. Plant species range from being almost self-sterile to being completely self-fertile. Cabbages can be self-fertilised, but they yield more seed when cross-fertilised by pollen from other plants, and cross-pollination by insects increases seed yield. Most modern varieties of oilseed rape, however, are almost completely self-fertile; self-pollination gives virtually the same seed yield as cross-pollination (Williams and others, 1986).

The flowers of cabbages and oilseed rape appear to be adapted to insect pollination, since they have several features that encourage insect visits, such as nectar and bright yellow petals. Many insects visit the flowers, particularly honeybees, bumblebees, solitary bees, flies, beetles, and thrips, and it is easy to see some of them pollinating. However, oilseed rape can also be wind-pollinated. In a glasshouse, plants were cross-pollinated when blown with a hair dryer to simulate wind (Eisikowitch, 1981). Many flowers are probably wind-pollinated in the field, since there is often a large amount of oilseed rape pollen in the air (Williams, 1984). Oilseed rape can also be "auto-pollinated" by pollen that falls off the long stamens onto the stigma within the same flower. Auto-pollination is probably enhanced when plants are shaken by the wind.

Oilseed rape yields well without insects. Insect pollination is therefore unlikely to give much increase in total seed yield, but it can increase a farmer's profit. Extra insects can speed up pollination, giving earlier, more uniform pod ripening and so increasing the efficiency of harvesting (p. 13; Williams and others, 1986).

The honeybee (*Apis mellifera*)

Honeybees are frequent visitors to flowering cabbages and oilseed rape. Most individuals suck up nectar, which they use as an energy source or carry back to the colony where it is converted into honey and stored. A few bees also collect pollen, which they pack into small pellets. These yellow pollen loads are carried on the bees' hindlegs back to the colony, where they are stored to provide the protein necessary for the growth of young bees. Cabbages are rarely allowed to flower, so they are of little significance to bees, but large areas of flowering oilseed rape are of enormous benefit. They provide a large amount of nectar and pollen, which helps to boost the size of the colony early in the bees' foraging season, so that there are plenty of worker bees to forage from plants that flower later in the season.

Although bees can fly as far as 12 km if there is no nectar available nearby, they usually forage within about 3 km of their colony. Beekeepers often move their beehives into fields of oilseed rape to ensure that as many bees forage there as possible. Honeybees usually suck nectar from the inner nectaries (fig. 14). They seem to ignore the outer nectaries, perhaps because these secrete less nectar. The bees usually enter the flowers from the front, and so transfer pollen from the body of the bee to the stigma, effecting pollination. Sometimes a bee stays outside the flower and feeds by pushing its tongue between the bases of the sepals and petals. This "base working" enables the bees to take nectar, but the flower is not pollinated.

Fig. 14. A honeybee visiting an oilseed rape flower from in front.

The honey from oilseed rape contains a particularly high concentration of glucose, so it crystallises or granulates rapidly, thickening or setting within a couple of weeks of being taken from the hive (Calder, 1986). The set honey is white with a mild, delicate flavour, although some say there is a faint cabbage taste. Beekeepers value oilseed rape because of its high yield, but the rapid crystallisation means extra work for them. Honey can only be extracted from the honeycomb while it is runny, so full honeycomb must be removed promptly and the honey extracted at once before it sets; it cannot just be left in the hive and extracted with all the other honey at the end of the honey season.

In the past, many honeybees were killed by insecticides used against seed weevil and pod midge because the sprays were applied while bees were still foraging on the crop. However, all pesticide users in the UK are now legally required to follow instructions relating to the use of each pesticide, some of which concern protection of bees. This often means waiting until flowering is over and warning local beekeepers at least two days in advance of spraying. Since many bees may still be visiting the crop, even when very few flowers are left (Free & Ferguson, 1980), farmers are advised to spray in the early morning or evening when fewer honeybees are flying. Bumblebees start earlier and finish later in the day than honeybees. It might be worth investigating the possibility that they may be working at times when spraying is recommended. Farmers can also help to protect beneficial insects by selecting insecticides which are of minimal hazard to bees. Fortunately, the development of new insecticides is increasingly taking honeybees into account. Some modern pyrethroid insecticides repel honeybees so that they return to the hive and stop foraging temporarily (Rieth & Levin, 1989); it may be possible to make insecticides more selective by such means. Of course it is not just honeybees that are beneficial. Bumblebees, solitary bees, and hoverflies also pollinate, and predatory insects, such as hoverfly larvae, flower bugs, and parasitic wasps, help to reduce the number of aphids and other pests.

The pollen beetle (*Meligethes aeneus*)

Several species of pollen beetle (*Meligethes* species) occur on cabbages and oilseed rape. They all look very similar. By far the commonest species is *M. aeneus*. The adults are 2–3 mm long, oval, and black with a metallic green sheen (pl. 1.5). As more and more oilseed rape has been grown, the numbers of pollen beetles have increased enormously. People have been surprised to find dozens of small black beetles landing on washing hanging out to dry or crawling out of cut flowers brought indoors. This is part of the normal behaviour of the beetles looking for flowers; it is just that the increase in numbers has made the beetles much more obvious.

The life cycle has been studied in detail by Williams (1978). Adults overwinter in the soil or amongst vegetation. In spring, they fly to flowering plants of a wide range of species and feed on pollen and nectar. When the females are ready to lay eggs, they fly to crucifers that are in bud or flowering. Here they feed on pollen in open flowers, if these are available (fig. 15); otherwise they bite holes in buds to reach pollen. When about to lay, they select a green bud 2–5 mm long and bite a hole through the base of the sepals and petals. One to three eggs are laid through the hole and glued to the stamens or style. Eggs hatch after 4–7 days. The larvae are creamy white with many small, dark brown plates, and up to 5 mm long (pl. 5.1). They feed on pollen for 9–13 days, often moving between flowers, and then drop to the ground to pupate in the soil (Williams & Free, 1978). The next generation of adults emerges after 14–18 days and flies to a wide range of flowering plants, including crucifers, to feed before overwintering. There is only one generation per year. Winter rape is usually less heavily infested than spring rape

Fig. 15. Adult pollen beetles in an oilseed rape flower

because it has passed the susceptible green-bud stage before large numbers of beetles arrive.

Traps have been used to investigate host-finding behaviour. Yellow traps catch more beetles than several other colours (Fritzche, 1957), and the addition of oilseed rape extract or 2-propenyl isothiocyanate increases catches further (Free & Williams, 1978). The beetles also appear to aggregate in response to the presence of others. Flowers marked with black dots to simulate beetles contained more beetles after 3 hours than unmarked flowers (Free & Williams, 1978). This could usefully be investigated further.

Adults and larvae cause little or no damage to open flowers, but the adults damage buds when they bite holes in them. The most severely damaged buds are liable to drop off without setting any seed, leaving a "podless stalk" (Williams & Free, 1978) (fig. 16). This symptom was once used as an index of pollen beetle damage, but it has been shown to be an unreliable one because many undamaged flowers fail to set seed and this also produces podless stalks. Oilseed rape plants have an enormous capacity to compensate for damage, particularly when this is to buds (see p. 13). Williams & Free (1979) found that as many as 40% of buds could be artificially removed from plants without causing any significant decrease in yield, so pollen beetles are unlikely to decrease yield unless the infestations are very heavy.

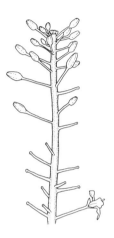

Fig. 16. Podless stalks on an oilseed rape plant.

Glucosinolates defend crucifers against insects, so when farmers began to grow double low varieties of oilseed rape with low levels of glucosinolate in the seed there was concern that these new varieties might be more vulnerable to insect damage. Surveys of pollen beetle infestation showed that although some varieties were regularly infested more heavily than others, there was no consistent difference between single low and double low varieties. The same was found for seed weevils, pod midges, and cabbage stem flea beetles (Williams, 1989). Chemical analyses so far suggest that the reduction in seed glucosinolates in double low varieties has not been clearly reflected in the glucosinolate level of flowering racemes or foliage, so that the insects may detect little if any difference (see Williams, 1989). The glucosinolate level in the seed is reflected in that of the young seedlings, however, so the seedlings of double low varieties may be more susceptible (Glen and others, 1990).

These rape pests are brassica specialists, which have overcome glucosinolate defences, so a decrease in glucosinolates seems unlikely to help them. Reduction of glucosinolate levels would reduce cues to many of their behaviours and so might even decrease, rather than increase, infestation. Perhaps we should expect any increase in damage as a result of decreased toxicity to come particularly from generalists and other herbivores not normally associated with crucifers (see p. 6). There is some evidence for this. Glen and others (1990) found that damage to a range of varieties of oilseed rape seedlings by a generalist feeder, the field slug (*Deroceras reticulatum* Müller), increased as the glucosinolate level decreased.

5 Identification

Introduction to the keys

These keys are intended to identify the insects commonly found on cabbages and oilseed rape grown in the United Kingdom, whether on farms, in gardens or in the wild. Insects that breed on the plant are usually identified to the level of species or genus (a group of species). Insects that are common, but more casually associated with the plant, are usually identified only as far as the family (a group of genera) or order (a group of families). In a book of this size, it is not practicable to allow for every species that could possibly be found on the plant. Many passers by will still key out into categories such as "other flies", but you must be prepared to leave specimens unidentified if they do not fit the key exactly. This may happen for rare species. It may also happen in the misleading case where winged adults are emerging from the soil, having bred on a previous different crop, or are spilling over from a different crop nearby. For example, cereal flies may be common in a rape field next to a cereal field. So if an abundant insect does not fit the key, it may not be associated with cabbages or rape; see what insects are common in nearby areas or were common on the previous crop.

There are several technical and common names for the immature stages of insects between the egg and adult stages. In this book, all active immature stages are referred to as larvae. Some larvae look very different from the adult and undergo a radical change of their body tissues during a resting stage called the pupa. This is well known in butterflies, where the larva is also known as a caterpillar and the pupa is also known as a chrysalis. The larvae of some other groups of insects are known as grubs or maggots. Some larvae (nymphs) look rather similar to the adult and gradually become more like the adult at each moult. There is no pupal stage. This type of life history is found in aphids.

Since eggs and pupae are very hard to identify, this key is restricted to larvae and adults. This means that, for virtually all species, the active stages are covered, while the inactive, non-feeding stages are not. Note that spiders, mites, centipedes, millipedes, woodlice and slugs are not insects and are not included!

To identify a specimen, you must have it in front of you. You will usually have to examine it under a hand lens or dissecting microscope with a good light source. This usually means you will need to keep the specimen completely still, by either killing or anaesthetising it (techniques p. 56). Before collecting immature stages, observe where on the plant you found them and what they were doing. This knowledge will be helpful for some parts of the key.

Start at Key I, the main key. Keys usually consist of numbered pairs of descriptions called couplets. Start at couplet 1 and decide which description fits your specimen. At the end of the description that fits, you will find either the number of the couplet or key to go to next, or the name of the species or group to which your specimen belongs. If a specimen fits neither of the descriptions in a couplet, you may have selected the wrong description at some stage. Try again, taking care over any difficult couplets. Higher magnification, better lighting, or consultation with a colleague may help. If the specimen still does not fit, it may be a species not included in the key because it is rarely found on the plant. It may be possible to identify it, at least to family, in Chinery (1976).

Plates 1 to 5 can be used for some species to confirm an identification from the key. They can also be used to suggest what a specimen might be, but they are not reliable for species identification because several other species may look identical to the species in the plate. Use them in conjunction with the keys.

The scientific names of species are given, with the common name as well when one exists for the species. A full scientific name consists of three parts, for example *Pieris brassicae* L. The first word, by convention printed in italics with a capital letter, is the genus, the group to which the species belongs. The second word, in italics without a capital letter, is the species epithet. The third word, not in italics, is the name of the author who first published a description of the species. Well known authors' names are often abbreviated; for example, Linnaeus is abbreviated to L. If the name of the genus has been changed since the author named the species, the author's name will appear in brackets. If the old genus name was well known, it is sometimes given in brackets after the new name, for example *Artogeia* (= *Pieris*) *rapae* (L.). The system is the same for plant names, except that if the author's name is in brackets, it is usually followed by the name of the person who changed the name of the genus; for example *Brassica nigra* (L.) Koch. In this book, authors' names are given in the keys. For species that are not included in the keys, the authors' names are given when the species is first mentioned in the text. When a scientific name is repeated, it is often abbreviated by giving the initial letter of the genus and dropping the author, as in *P. brassicae*.

It is often useful to know the name of the insect order (usually ending in -ptera), family (ending in -idae), sub-family (ending in -inae) or other group to which a species belongs, particularly when trying to look it up. This information is given either part-way through the key or in brackets after the species names.

The lengths given are body lengths, unless stated otherwise, from the front of the head to the tip of the abdomen, excluding any antennae, wings, legs, hairs or snout (relevant in adult weevils: see Key III). These indicate typical sizes. However, unusually large or small individuals may occur, so unless the sizes come first in a description, they should be used only to confirm identifications and not relied upon to separate species.

* = major pest species or groups on cabbages or oilseed rape (see table 4).
+ = beneficial species or groups.

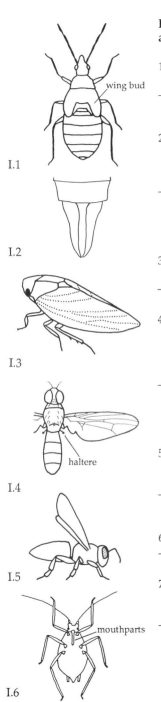

I.1

I.2

I.3

I.4

I.5

I.6

wing bud

haltere

mouthparts

Key I. The major groups of insects found on cabbages and oilseed rape

1   With wings; sometimes hard, opaque, wing covers hide
    flexible transparent hindwings. Flying adults   2
–   Without wings or with wing buds (I.1). Larvae and non-
    flying adults   Key II

2   Forewings hard or leathery all over, opaque and usually
    dark, forming wing cases; hindwings flexible and
    transparent, usually completely hidden by the wing
    cases   3
–   Forewings at least partly soft and flexible, not forming
    wing cases; forewings and hindwings (if present)
    appearing at least partly clear or translucent, unless
    obscured by pigmented scales or white wax   5

3   With forceps on end of abdomen (I.2)
        common earwig *Forficula auricularia* L. (Dermaptera)
–   Without forceps on end of abdomen   4

4   Hard or leathery forewings with veins, meeting at an
    angle (roof-wise) down the middle of the back (I.3);
    usually green or brown
                leaf hoppers, plant hoppers and their allies
                (Hemiptera: Homoptera) (see Chinery, 1976)
–   Hard or leathery forewings without veins, though
    sometimes with many parallel grooves; forewings not
    meeting at an angle (roof-wise) down the middle of the
    back (pl. 1.1–1.8). Beetles (Coleoptera)   Key III

5   One pair of wings; a club-shaped haltere, often yellow or
    whitish, immediately behind each wing (I.4). Flies
    (Diptera)   Key IV
–   Two pairs of wings; sometimes the hindwings are small
    and hidden beneath the forewings   6

6   Wings covered with pigmented scales or white wax   7
–   Wings without scales or wax   8

7   Wings white all over; body length 1-3 mm; on lower
    surfaces of leaves (pl. 4.1)   *cabbage whitefly
        Aleyrodes proletella (L.) (Hemiptera: Aleyrodidae)
–   Wings not white all over; body length more than 6 mm.
    Butterflies and moths (Lepidoptera)   Key V

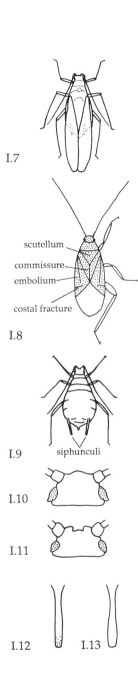

scutellum

commissure

embolium

costal fracture

I.7

I.8

I.9   siphunculi

I.10

I.11

I.12   I.13

8   Abdomen not strongly narrowed into a waist behind the
     thorax when viewed from the side   9
–   Abdomen strongly narrowed into a waist behind the
     thorax when viewed from the side (I.5); sometimes the
     waist is hidden by hairs; antennae often elbowed.
     Ants, bees, wasps, and parasitic wasps (Hymenoptera:
     Apocrita)   34

9   Narrow, elongate, sucking mouthparts (rostrum) visible
     beneath head (I.6). Aphids, hoppers and bugs
     (Hemiptera)   10
–   Sucking mouthparts not visible (sucking mouthparts
     retracted or only biting mouthparts present); often the
     mouthparts are within or at the end of a projecting part
     of the head   19

10   Forewings not overlapping when folded, entirely soft or
     entirely stiff (I.7). Hemiptera: Homoptera   11
–   Forewings overlapping flat over the body when folded,
     partly hard or leathery but with soft tips (I.8).
     Hemiptera: Heteroptera   14

11   Abdomen with a pair of tubes (siphunculi) at the end
     (I.9); head and thorax dark brown to black; wings clear
     with dark veins; less than 3 mm. Aphids   12
–   Abdomen without a pair of tubes at the end; body
     usually green or brown; wings usually with some
     coloured markings; jumping   leaf hoppers,
               plant hoppers and their allies (see Chinery, 1976)

12   Head at base of antennae not swollen (I.10); head,
     thorax and antennae blackish; abdomen green with
     black markings; siphunculi not extending as far as the
     end of the abdomen   *cabbage aphid
     or mealy cabbage aphid *Brevicoryne brassicae* (L.)
–   Head at base of antennae swollen and projecting into
     the space between the antennae (I.11)   13

13   Siphunculi dark with darker tips, only slightly, if at all,
     swollen in the outer half (I.12), usually long enough to
     project beyond the end of the abdomen; ground colour
     of abdomen greenish yellow to green
               *peach-potato aphid *Myzus persicae* (Sulzer)
–   Siphunculi dark without darker tips, distinctly swollen
     in the outer half (I.13), usually not long enough to
     project beyond the end of the abdomen; ground colour
     of abdomen dark green
               shallot aphid *Myzus ascalonicus* Doncaster

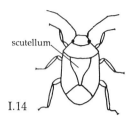

I.14

14   Scutellum longer than the distance from the rear end of
      the scutellum to the end of the body (I.14); antennae
      with 5 segments. Shieldbugs (Pentatomidae)   15
 −   Scutellum shorter than the distance from the rear end of
      the scutellum to the end of the body (I.8); antennae with
      four segments   16

15   Body metallic blue-black or green-black with red, yellow
      or white markings; antennae black; upper surface of
      head with a slightly raised rim; 6-8 mm (pl. 3.7)
                                  brassica bug *Eurydema oleracea* (L.)
 −   Not like this   other shieldbugs (Pentatomidae)
                                   (see Southwood & Leston, 1959)

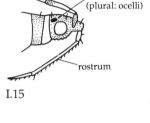

I.15

16   Ocelli present (I.15); rostrum 3-segmented, curved and
      not pressed against the body when the insect is resting
      (I.15); forewings with costal fracture and embolium (I.8);
      less than 5 mm; mainly predatory; with a combination
      of black, dark brown or reddish markings. Flower bugs
      (Anthocorinae)   17
 −   Not like this   18

17   Forewings entirely shining, with brown tips and black
      markings; segment behind head (prothorax) entirely
      black; 4 mm (pl. 3.6)
                     +common flower bug *Anthocoris nemorum* (F.)
 −   Not like this   +other flower bugs
                                   (see Southwood & Leston, 1959)

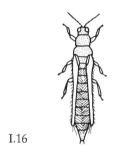

I.16

18   Ocelli absent; rostrum 4-segmented, pressed flat against
      the body when at rest; forewings with costal fracture
      (I.8); scutellum shorter than commissure (I.8); mainly
      plant feeding; usually green or brown
          capsid bugs (Miridae) (see Southwood & Leston, 1959)
 −   Not like this   other bugs (Hemiptera:
          Heteroptera) (see Southwood & Leston, 1959)

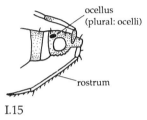

wing fringes not shown
I.17

19   Less than 3 mm; wings strap-like and fringed with hairs,
      on top of abdomen when at rest (I.16, pl. 4.4). Thrips
      (Thysanoptera)   20

      (Thrips will have to be mounted on microscope slides and examined
      under a compound microscope in order to see most of the characters
      described below. See p. 56 for details.)
 −   More than 3 mm   31

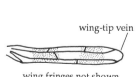

wing fringes not shown
I.18

20   Forewings broad, with cross-veins (I.17). Aeolothripidae

                                                          21
 −   Forewings narrow, without cross-veins (I.18).
      Thripidae   23

21 Forewings pigmented black/grey/brown along the
    whole wing, slightly paler at base
    Melanthrips fuscus (Sulzer)
 − Forewings with 2 black/grey bands (I.17); live
    specimens have the wings along the body and so appear
    to be striped black and white   22

22 Wing-tip vein (I.17) darker than the surrounding
    membrane   *Aeolothrips tenuicornis* Bagnall
 − Wing-tip vein as pale as the surrounding membrane
                                Aeolothrips intermedius Bagnall

23 Upper side of head, excluding antennae, at least as long
    as it is wide (I.19)   cereal thrips *Limothrips* species
 − Upper side of head, excluding antennae, wider than it is
    long (I.20)   24

I.19

24 Forewing with an interrupted row of bristles on the first
    vein (I.18)   25
 − Forewing with a complete row of bristles on the first
    vein   *Frankliniella* species

25 Antennae with 7 segments   26
 − Antennae with 8 segments; body black or dark brown   30

I.20

26 All abdominal sternites (underside plates) with bristles
    confined to the hind margin   27
 − Some abdominal sternites with some bristles away from
    the hind margin (I.21)   28

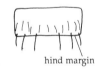

27 Body yellow, cream or brown; antennal segments 3 and
    4 (counting away from the head) of uniform colour;
    abdominal tergite (upperside plate) 8 with a complete,
    uninterrupted comb along the hind margin (I.22)
                                onion thrips *Thrips tabaci* Lindeman
 − Not like this
                        other *Thrips* species (see Mound and others, 1976)

I.21   hind margin

28 Forewings short, not reaching as far as abdominal
    segment 5 (I.22) when folded along the back; body
    brown to black   cabbage thrips (short- winged forms)
                                        Thrips anguisticeps Uzel
 − Forewings long, reaching beyond abdominal segment 5
    when folded along the back   29

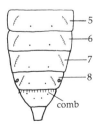

comb

I.22

29 Body brown to black; forewings with 6–11 bristles on
    the outer half of the first vein (I.18) (pl. 4.4)
                            cabbage thrips *Thrips anguisticeps* Uzel
 − Not like this
                        other *Thrips* species (see Mound and others, 1976)

30   Forewing with 3 bristles on the outer half of the first
     vein   *Thrips vulgatissimus* Haliday
–   Forewing with more than 3 bristles on the outer half of
     the first vein   *Thrips atratus* Haliday

31   Wings with a network of veins and many cross-veins
     (I.23); body usually green or brown
     lacewings (Neuroptera) and scorpion flies (Mecoptera)
–   Wings with few cross-veins (I.24); body usually with
     some combination of black, yellow or orange markings.
     Sawflies (Hymenoptera: Symphyta)   32

I.23

32   Antennae with 9 segments; forewing with 5-sided
     central cell (area surrounded by veins) (I.24); abdomen
     all black (female) or black with a red band on segments
     2–6 counting away from the thorax (male); stigma (I.24),
     antennae and upper surface of head entirely black;
     9–12 mm   *Tenthredo atra* L.
–   Not like this   33

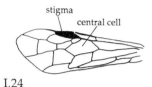

stigma

central cell

I.24

33   Forewing with 4-sided central cell (I.25); thorax black
     and yellow-orange; abdomen yellow-orange apart from
     some black next to the thorax; stigma and upper surface
     of head and antennae black; antennae appearing to have
     from 9–11 segments; 6–9 mm
                          turnip sawfly *Athalia rosae* (L.)

(This species has been a major pest in England in the past,
particularly on turnips, but it is now rare. It is a pest of oilseed rape
on the Continent and so it could perhaps spread here again. If you
find it, have the identification checked, perhaps at your local
museum. Your nearest ADAS office (Agricultural Development and
Advisory Service) may be interested if you find this species.)

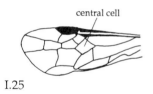

central cell

I.25

–   Not like this   other sawflies (Hymenoptera: Symphyta)
                       (Quinlan & Gauld, 1981; Benson, 1952–1958)

34   Forewing with 1 or no enclosed cells (areas of the wing
     completely enclosed by veins or by veins and the wing
     base)   +parasitic wasps
           (Chalcidoidea: Eulophidae and other families)

(If it has been reared from cocoons beside *Pieris brassicae* larvae, it
may be the hyperparasitoid *Tetrastichus galactopus* (Ratzeburg).
See p. 9 and fig. 6. If work is to be published, identification must be
checked by an expert.)

–   Forewing with two or more enclosed cells   35

35   Antennae with 16 or more segments. Parasitic wasps
     (Ichneumonoidea)   36
–   Antennae with fewer than 16 segments   37

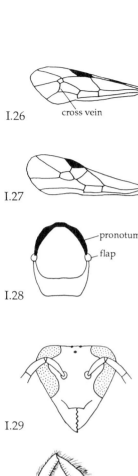

I.26   cross vein

I.27

36 Forewing with a cross-vein in the lower outer quarter
    (I.26)   +parasitic wasps (Ichneumonidae)
    (If it has been reared from cocoons beside *Pieris brassicae* larvae, it
    may be the hyperparasitoid *Lysibia nana* (Gravenhorst). See p. 9. If
    work is to be published, identification must be checked by an
    expert.)

–   Forewing without a cross-vein in the lower outer
    quarter (I.27)   +parasitic wasps (Braconidae)
    (If it has been reared from cocoons beside *Pieris brassicae* larvae, it
    may be the parasitoid *Apanteles glomeratus* (L.). See p. 9 and figs 4
    and 5. If work is to be published, identification must be checked by
    an expert.)

I.28 — pronotum, flap

37 Forewing with 5 or fewer enclosed cells   38
–   Forewing with 6 or more enclosed cells   39

38 Body bright metallic blue, green and sometimes red
                    ruby-tailed wasps (Chrysididae)
–   Not bright metallic; shiny black or brown
                              +other parasitic wasps

I.29

39 Pronotum extending backwards as far as the flaps over
    the wing bases (view the thorax from above) (I.28)   40
–   Pronotum not extending backwards as far as the flaps over
    the wing bases, or thorax too hairy for this to be seen   42

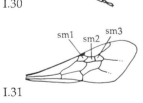

I.30   tarsus

40 Margin of eye indented next to antenna (I.29)
    social wasps (Vespidae) and potter wasps (Eumenidae)
–   Margin of eye not indented by antenna   41

41 Antennae elbowed, with a sudden angled bend; body
    uniform brown to black   ants (Formicidae)
–   Antennae straight or curved; body usually black with
    orange markings; long legs used for handling prey
                    +spider-hunting wasps (Pompilidae)

I.31   sm1 sm2 sm3

42 First segment of hind tarsus (counting away from the
    body) as broad as the next tarsal segment, often with
    short hairs, but never very hairy; body never very hairy
              +solitary wasps (Sphecidae) (see Yeo & Corbet, 1983)
–   First segment of hind tarsus much broader than the
    other tarsal segments (I.30) and often very hairy,
    sometimes with a lump of pollen attached; body often
    very hairy. Bees   43

I.32   mandible

43 Forewings with 3 sm (sub-marginal) cells, cell sm2 the
    same area or larger than cell sm3 (I.31); eyes not reaching
    down to mandibles (view from the side) (I.32)   44
–   Not like this   +solitary bees (Apidae)

44   Eyes smooth; body densely hairy and fat, the hair
      usually appearing brown all over, or black with various
      white, yellow or red stripes   +bumblebees
      (Apidae: *Bombus* species and *Psithyrus* species)
      (see Prŷs-Jones & Corbet, 1991)
–   Eyes hairy; body with sparse yellow and black hairs;
      abdomen black or brown, usually with some orange
      stripes across the width   +honeybee *Apis mellifera* L.

Key II. Larvae and non-flying adults

1   Living in or on another insect; parasites and parasitoids
                              +parasitic wasp, fly or beetle larvae
      (These are best identified after rearing them to the adult stage. See
      p. 57 for advice on rearing. Key I will identify some parasite and
      parasitoid adults to family. Information on parasites and parasitoids
      is available for some larval and pupal hosts: *Pieris brassicae*
      (Littlewood, 1988); *Artogeia rapae* (Richards, 1940); *Delia* species
      (Wishart and others, 1957); and *Meligethes* species (Osborne, 1960).)
–   Not living in or on another insect; herbivores or
      predators   2

2   On the plant surface above soil level, usually on leaves
      or in flowers   3
–   Grubs, maggots and caterpillars found inside stems, leaf
      stalks, leaves or pods, or living below soil level   46

3   Live specimen does not walk, crawl or curl up, even if
      prodded; firmly fixed to one spot on the plant   4
–   Live specimen can walk, crawl or curl up (feeding
      aphids and some caterpillars may not react at once if
      prodded, but will eventually)   5
      (If you cannot use this couplet because your specimen is dead and
      you did not observe it when it was alive, first see if it fits couplet 4
      and then, if it does not, move to couplet 5)

4   Oval scale, flat against the plant surface; 1 mm.
      Hemiptera: Homoptera   *cabbage whitefly larvae
                    Aleyrodes proletella (L.) (Aleyrodidae)
      Not like this; without legs; surface usually hard.
      Eggs or pupae   not covered by this key
      (Clusters of about 30–100 skittle-shaped yellow eggs on the
      underside of leaves are likely to be those of the large white butterfly
      Pieris brassicae. See fig. 3 and p. 8.)

5   Antennae at least as long as the head; some larvae,
      wingless adults or short-winged (non-flying) adults   6
–   Antennae absent or not as long as the head; grubs,
      maggots or caterpillars   15

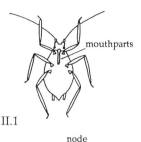

II.1

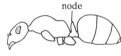

II.2

II.3

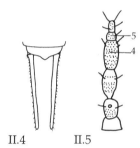

II.4   II.5

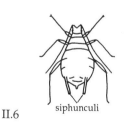

II.6

II.7

6   Sucking mouthparts not visible (sucking mouthparts retracted or only biting mouthparts present); often the mouthparts are within or on the end of a projecting part of the head   7

–   Narrow, elongate sucking mouthparts (rostrum) visible beneath head (II.1). Aphids and bugs (Hemiptera)   12

7   Length more than 2 mm; biting mouthparts visible   8

–   Length 2 mm or less; without biting mouthparts; sucking mouthparts retracted within head. Thrips (Thysanoptera) 10

8   Abdomen strongly narrowed into a waist behind the thorax when viewed from the side; waist formed from one or two segments each with an expanded node or scale (II.2); mandibles (jaws) short, not projecting forwards; active, fast-moving   ant adults
(Hymenoptera: Formicidae) (see Rotheray, 1989)

–   Abdomen not strongly narrowed into a waist   9

9   Mandibles long, projecting forwards (II.3); without forceps on the end of the abdomen
+lacewing larvae (Neuroptera) (see Rotheray, 1989)

–   Mandibles short, not projecting forwards; with forceps, which are often simple and straight, on the end of the abdomen (II.4, I.2)   common earwig larvae
Forficula auricularia L. (Dermaptera)

10   With small non-functional wings, appearing to be wing buds; body black or dark brown
cabbage thrips adults (short-winged forms)
Thrips angusticeps Uzel (Thripidae)

–   Without wings or wing buds. Thrips larvae (Thysanoptera)   11

11   Antennal segment 5 less than half the length of segment 4 (II.5)   Thripidae

–   Antennal segment 5 at least half the length of segment 4
Aeolothripidae

12   Abdomen with a pair of tubes (siphunculi) near the end of the abdomen (II.6). Aphid larvae and wingless adults (Hemiptera: Homoptera)   13

–   Abdomen without a pair of tubes near the end of the abdomen   bug larvae and short-winged adults
(Hemiptera: Heteroptera)(see Southwood & Leston, 1959)

13   Head at base of antennae not swollen (II.7); body greyish green with black markings; covered in a fine white dust; usually in dense colonies wet with honeydew (pl. 4.2)   *cabbage aphid
or mealy cabbage aphid *Brevicoryne brassicae* (L.)

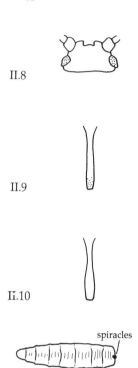

II.8

II.9

II.10

spiracles

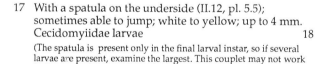

II.11

spatula

II.12

– Head at base of antennae swollen and projecting into the space between the antennae (II.8); usually not in dense colonies   14

14 Siphunculi with dark tips, only slightly, if at all, swollen in the outer half (II.9); body light green to yellow or pinkish (pl. 4.3)

*peach-potato aphid *Myzus persicae* (Sulzer)

– Siphunculi without dark tips, distinctly swollen in the outer half (II.10); body greyish yellow to greenish brown   shallot aphid *Myzus ascalonicus* Doncaster

15 Without legs on the 3 segments behind the head; head usually not distinct and not hardened and darkened above (pl. 5.5). Fly larvae (Diptera)   16

– With legs on the 3 segments behind the head; head always distinct and hardened, often darkened above (pl. 2.1–2.8 and 5.1)   20

16 Rear end of upperside of body with two spiracles (breathing holes) united in a single brown or black respiratory tube (II.11); flattened from top to bottom and "slug-like"; predatory on aphids; often with some coloured markings

+hoverfly larvae (Syrphidae) (see Rotheray, 1989)

(*Syrphus ribesii* (L.) is common. The last instar is off-white with a pinkish stripe and surrounding white markings down the middle of the back, about 12 mm, and with a short, light brown respiratory tube more than 0.55 mm wide at the tip.)

– Rear end of body with 2 separate spiracles, which may project slightly   17

17 With a spatula on the underside (II.12, pl. 5.5); sometimes able to jump; white to yellow; up to 4 mm. Cecidomyiidae larvae   18

(The spatula is present only in the final larval instar, so if several larvae are present, examine the largest. This couplet may not work for larvae under about 2 mm long.)

– Without a spatula; unable to jump   19

18 Able to jump; yellowish white when young, lemon yellow later; on leaves, leaf stalks and stems or in flowers; stunts leaves, stems and flowers; up to 4 mm; May–October

swede midge *Contarinia nasturtii* (Kieffer)

– Non-jumping; whitish at all ages; in flowers; stunts the flower, preventing seed development; gregarious; up to 3 mm   brassica flower midge *Gephyraulus raphanistri* (Kieffer)

(Identification could be checked by trying to rear the larvae to adults and then using Key IV. See p. 57 for advice on rearing. For certainty, have the identification checked by an expert.)

PLATE 1

1. *Phyllotreta atra*
   (Flea beetle)

2. *Phyllotreta nemorum*
   (Large striped flea beetle)

3. *Phyllotreta undulata*
   (Small striped flea beetle)

4. *Psylliodes chrysocephala*
   (Cabbage stem flea beetle)

5. *Meligethes aeneus*
   (Pollen beetle)

6. *Ceutorhynchus assimilis*
   (Cabbage seed weevil)

7. *Ceutorhynchus quadridens*
   (Cabbage stem weevil)

8. *Ceutorhynchus picitarsis*
   (Rape winter stem weevil)

(x 8 natural size)

1

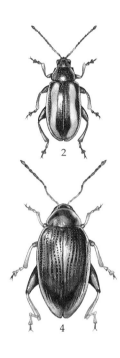

2

3

4

5

6

7

8

PLATE 2

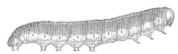

1. *Pieris brassicae*
   (Large white butterfly)
   larva

2. *Artogeia rapae*
   (Small white butterfly)
   larva

3. *Artogeia napi*
   (Green-veined white butterfly)
   larva

4. *Autographa gamma*
   (Silver Y moth)
   larva

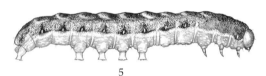

5. *Mamestra brassicae*
   (Cabbage moth)
   larva

6. *Melanchra persicariae*
   (Dot moth)
   larva

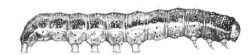

7. *Lacanobia oleracea*
   (Tomato moth or bright-line
   brown-eye moth)
   larva, dark form

8. *Agrotis segetum*
   (Turnip moth)
   larva

(x 1.75 natural size)

PLATE 3

1. *Plutella xylostella*
    (Diamond-back moth)
    larva

2. *Evergestis forficalis*
    (Garden pebble moth)
    larva

3. *Phytomyza rufipes*
    (Cabbage leaf miner)

4. *Delia radicum*
    (Cabbage root fly)
    female

5. *Scathophaga stercoraria*
    (Yellow dung-fly)

6. *Anthocoris nemorum*
    (Common flower bug)

7. *Eurydema oleracea*
    (Brassica bug)

(1–2 and 4–7 are
x 3.5 natural size,
3 is x 5 natural size)

PLATE 4

1. *Aleyrodes proletella*
   (Cabbage whitefly)

2. *Brevicoryne brassicae*
   (Cabbage aphid)
   wingless adult female

3. *Myzus persicae*
   (Peach-potato aphid)
   wingless adult female

4. *Thrips angusticeps*
   (Cabbage thrips)

5. *Dasineura brassicae*
   (Brassica pod midge)
   female with ovipositor
   extended

6. *Scaptomyza flava*

(1 is x 25 natural size,
2–3 and 5–6 are
x 15 natural size,
4 is x 45 natural size)

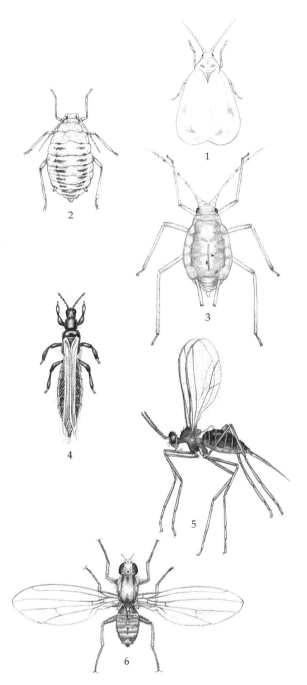

PLATE 5

1. *Meligethes aeneus*
   (Pollen beetle)
   larva, 2nd instar

2. *Ceutorhynchus assimilis*
   (Cabbage seed weevil)
   larva

3. *Psylliodes chrysocephala*
   (Cabbage stem flea beetle)
   larva

4. *Phyllotreta nemorum*
   (Large striped flea beetle)
   larva

5. *Dasineura brassicae*
   (Brassica pod midge)
   larva, underside

6. *Delia radicum*
   (Cabbage root fly)
   larva

(1–2 are x 15 natural size,
3–4 and 6 are
x 11 natural size,
5 is x 35 natural size)

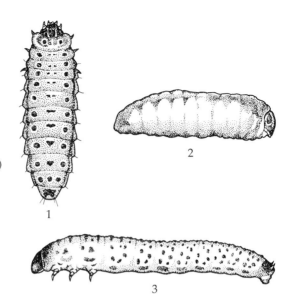

1

2

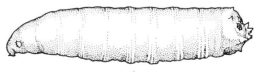

3

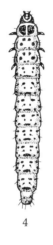

4

5

6

PLATE 6

1. A flowering stem of a brassica

2. Adult pollen beetles
   Meligethes in a flower

3. A honeybee *Apis mellifera*
   visiting a flower

4. A cabbage seed weevil
   Ceutorhynchus assimilis
   on a pod

5. Eggs laid by a large
   white butterfly *Pieris
   brassicae* on a leaf

6. A cabbage root fly
   Delia radicum laying eggs
   at the base of a stem

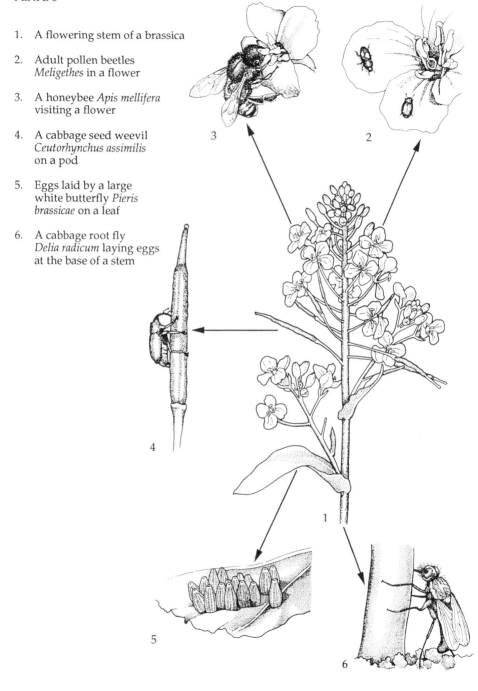

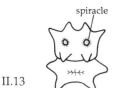

II.13

19 Grey or greyish brown; head retractable into the body; rear end of body with 2 spiracles (breathing holes) surrounded by 6 fleshy pointed lobes (II.13); feeding on young plants; up to 40 mm   leatherjackets (Tipulidae)
– Not like this   other fly larvae (Diptera)

20 With distinct prolegs (II.14)   21
– Without distinct prolegs. Beetle larvae (Coleoptera)   40

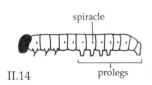

II.14

21 Three or fewer pairs of prolegs. Moth larvae (Lepidoptera: Heterocera) (part)   22
    (Identifications of moth caterpillars can be checked by rearing the adults and then using Key V. See p. 57 for advice on rearing.)
– Five or more pairs of prolegs   23

22 Two pairs of prolegs; body colour variable, greyish brown to yellowish green, paler below the line of spiracles (breathing holes, II.14); up to 25 mm; June–October   garden carpet moth
                    Xanthorhoe fluctuata (L.) (Geometridae)
– Three pairs of prolegs; usually greenish with white/yellow stripes along the body and with a green head, but can also be dark green to black with a paler underside and black head; up to 40 mm; May–September (pl. 2.4)
                    silver Y moth *Autographa gamma* (L.) (Noctuidae)

23 Eight pairs of prolegs. Sawfly larvae (Hymenoptera: Symphyta)   24
    (Sawfly larvae are not easy to separate. Couplets 24 and 25 apply to older larvae. For certainty you should have the identification checked by an expert or rear the larvae to adults, which are easier to identify, and use Key I. See p. 57 for advice on rearing.)
– Five pairs of prolegs. Butterfly and moth larvae (Lepidoptera) (part)   26
    (Identifications of butterfly and moth caterpillars can be checked by rearing the adults and then using Key V. See p. 57 for advice on rearing.)

II.15

24 Abdomen green with a dark stripe along the top; prolegs without hairs on the outer side; abdominal segments with 7 transverse folds and small white warts on the 2nd and 4th folds from the front (II.15); head with patches composed of black-brown dots; up to 23 mm
                    Tenthredo atra L.
– Not like this   25

II.16

25 Abdomen dark green to grey-black above with pale side stripes, greyish underneath, slightly flattened from top to bottom, frequently with darker markings just above the side stripes and a darker stripe along the top; prolegs with hairs on the outer side; abdominal segments with 6 transverse folds without warts or hairs; head shining black; up to 18 mm; mines leaves at first and then skeletonises them
          turnip sawfly *Athalia rosae* (L.)

(This species has been a major pest in England in the past, particularly on turnips, but it is now rare. It is a pest of oilseed rape on the Continent and so it could perhaps spread here again. If you find it, have the identification checked, perhaps at your local museum, or try rearing it to the adult stage and identifying it with Key I. See p. 57 for advice on rearing. Your nearest ADAS office (Agricultural Development and Advisory Service) may be interested if you find this species.)

II.17

–   Not like this
          other sawfly larvae (Hymenoptera: Symphyta)

26 Tips of prolegs on segments 6–9 (counting from behind the head) with hooks forming a complete circle or ellipse (use a microscope with high magnification) (II.16); up to 22 mm   27
–   Tips of prolegs on segments 6–9 with hooks forming a straight or curved row, never more than three quarters of a circle, along the inner edge (II.17, II.21); up to 50 mm   29
(One species that may appear intermediate is keyed out both ways.)

II.18

27 First segment behind the head with 2 hairs immediately in front of each spiracle (breathing hole, II.14) (II.18); yellowish green to greyish green, usually with a dark green stripe along the top of the body and a whitish side stripe; segments 4–11 with a single black spot immediately above each spiracle that is at least twice as wide as the spiracle; up to 22 mm; June–October; often webs leaves together (pl. 3.2)   *garden pebble moth
          Evergestis forficalis (L.) (Pyralidae)
–   First segment behind the head with 3 hairs immediately in front of each spiracle (II.19, II.20); up to 12 mm   28

II.19

28 Segments 6–9 with the two hairs below the spiracle spaced apart; usually greenish, sometimes pink or brown; up to 12 mm; May–October; mines leaves at first then feeds on the lower surface, often leaving the upper surface intact (pl. 3.1)
          *diamond-back moth *Plutella xylostella* (L.)
          (Yponomeutidae)

(This species used to be known as *Plutella maculipennis* (Curt.).)

II.20

–   Segments 6–9 with the two hairs below the spiracle close together on a small plate; greyish, bluish white, brownish cream or grey-green; up to 12 mm; September–June; mines leaves at first then feeds on the surface, binding leaves together with silk   flax tortrix moth
          Cnephasia interjectana (Haworth) (Tortricidae)

II.21

29 Tips of prolegs on segments 6–9 with the row of hooks a mixture of 2 or more sizes (use a microscope with high magnification and good lighting) (II.21); ground colour and markings green, yellow, white or black, never brown   30

–   Tips of prolegs on segments 6–9 with the row of hooks of uniform size or gradually changing in size along the row (II.17); fully grown larva rolls up when touched. Noctuidae (moths)   33
(Couplets 33–39 may not work for caterpillars under 20 mm long.)

30 Segments 4–11 with a single black spot above each spiracle (breathing hole, II.14); spots at least twice as wide as the spiracles; yellowish green to greyish green, usually with a dark green stripe along the top of the body and a whitish side stripe; up to 22 mm; June–October; often webs leaves together (pl. 3.2)
                *garden pebble moth *Evergestis forficalis* (L.)
                                                         (Pyralidae)

–   Segments 4–11 with many black spots or dots above each spiracle (pl. 2.1–2.3). Pieridae (butterflies)   31
(Couplets 31–32 may not work for caterpillars under 10 mm long.)

31 Upper surface with some black spots larger than the spiracles; ground colour bluish green to yellowish with a yellow stripe along the top of the body; up to 42 mm; June–October; often in groups and with a strong cabbage smell; skeletonises leaves; large solitary inactive individuals may have been parasitised by a parasitic wasp (fig. 5) (pl. 2.1)   *large white
                or cabbage white butterfly *Pieris brassicae* (L.)

–   Upper surface with none of the black dots larger than the spiracles; ground colour green; usually found singly (pl. 2.2, 2.3)   32

32 Yellow line along the top of the body; black dots present below the line of spiracles; up to 29 mm; May–August (pl. 2.2)   *small white butterfly *Artogeia rapae* (L.)
(This species is also widely known as *Pieris rapae* (L.).)

–   Green line along the top of the body, similar in colour to the rest of the upper surface; black dots absent below the line of spiracles; up to 26 mm; June–September (pl. 2.3)   green-veined white butterfly *Artogeia napi* (L.)
(This species is also widely known as *Pieris napi* (L.).)

33 In the soil or at soil level by day; feeding by night on roots and sometimes on seedlings, stems or low leaves. Cutworms   34

–   On the leaves by day   38

II.22

34 Brownish or greenish with a series of brownish black dashes down either side of the middle of the back and a paler mark along the outside edge of each dash   35

– Without a twin series of brownish black dashes on the back; middle segments with 4 small black-haired spots in the middle of the back, sometimes the front two spots very faint; brownish to greyish   36

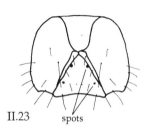

II.23   spots

35 Greyish brown with a reddish tinge along the back; dashes on the 11th segment joined by a band at the rear of the segment; up to 45 mm; June–November
                    pearly underwing *Peridroma saucia* (Hübner)

– Yellowish brown to greenish grey or bright green (very variable); dashes on the 11th segment not joined together; up to 50 mm; September–April
                    large yellow underwing *Noctua pronuba* (L.)

36 Upper surface covered with round, dark spots, many of which are larger than the dark areas immediately surrounding the bases of the uppermost hairs (use a microscope with high power and good lighting) (II.22); plate above the mouth without dark spots on the sides (II.23); up to 45 mm; April–July   dark sword-grass
          or dark dart *Agrotis ipsilon* (Hufnagel)

– Upper surface covered with many-sided platelets smaller than the dark areas immediately surrounding the bases of the uppermost hairs (II.24)   37

II.24

37 Spiracles (breathing holes, II.14) smaller than the wart-like dots immediately behind; plate above the mouth usually with 2 pairs of dark spots on the sides (II.23); without pear-shaped markings down the back; up to 47 mm; January–December (pl. 2.8)
                    turnip moth *Agrotis segetum* (D. & S.)

– Spiracles at least as large as the wart-like dots immediately behind; plate above the mouth without dark spots on the sides; often with a series of dark brown pear-shaped markings down the middle of the back, narrow end pointing backwards; up to 40 mm; January–December
          heart and dart moth *Agrotis exclamationis* (L.)

38 Segment behind the head with a brownish to greenish plate with 3 parallel whitish lines; end of the body with an angled hump; green, grey or brown with backward-pointing V-shaped marks; up to 45 mm; August–October (pl. 2.6)
          dot moth *Melanchra persicariae* (L.)

– Segment behind the head without 3 parallel whitish lines; end of body without an angled hump   39

39 Without white spots on the back; variable colouring green to brown, grey or black, with or without markings; usually with a yellowish or greenish white line next to the line of the spiracles (II.14), separating a darker upperside from a paler underside; up to 45 mm; June–October (pl. 2.5)   *cabbage moth Mamestra brassicae* (L.)

– Over 30 white spots per segment on the back; dark form is brownish to pinkish with a yellowish line next to the spiracles; light form is greenish with a yellow line next to the spiracles; up to 40 mm; July–September (pl. 2.7)
                    tomato moth or bright-line brown-eye
                    Lacanobia oleracea (L.)

40 In flowers; head brown-black; rest of body creamy white with, on top of each segment, a dark brown rounded plate each side and 1 or 2 smaller plates in between (pl. 5.1) (individuals 3–5 mm long with these plates faint to absent are prepupae); less than 5 mm. Pollen beetles (Nitidulidae: *Meligethes* species)   41
(Couplet 41 separates larvae, but not prepupae.)

– On leaves; not as above   42

41 Pale brown cuticular spots more or less evenly distributed on top of the thorax and abdomen (pl. 5.1)
                    pollen beetle Meligethes aeneus (F.)

– Pale brown cuticular spots present on top of the thorax and abdomen, but absent in a band between and immediately behind the 2 large plates on each segment (II.25)   pollen beetle *Meligethes viridescens* (F.)

II.25

42 Body arched above, flat beneath; legs elongate; ground colour of abdomen grey, blue-grey or black
                    +ladybirds (Coccinellidae)
                    (see Majerus & Kearns, 1989)

– not like this   43

43 Abdomen with a distinct pair of prongs at the end (II.26) 44

– Abdomen without a distinct pair of prongs at the end   45

II.26

44 Body, apart from head, dirty white with brown plates on the upper surface; up to 10 mm; November–April
                    turnip mud beetle *Helophorus* species (Hydrophilidae)

– Body, apart from head, pale, narrow and without plates; up to 6 mm   +rove beetle *Tachyporus* species (Staphylinidae)

45 Head and segment immediately behind head dark brown or black; rest of body yellowish brown with black spots and a row of brown projecting spots down each side; hairy; May–September; up to 6 mm
                    mustard beetle or watercress beetle
                    Phaedon cochleariae (F.) (Chrysomelidae)

– Not like this   other beetle larvae (Coleoptera)

46 Head distinct (pl. 2.8 and 5.2–5.4), hardened and
darkened above; with or without legs on the 3 segments
behind the head   47
–   Head not distinct (pl. 5.5, 5.6), not hardened and
darkened above; without legs on the 3 segments behind
the head. Fly larvae (Diptera)   49

47 With legs on the 3 segments behind the head
(pl. 5.3, 5.4)   48
–   Without legs; creamy white humped body with a
brownish head (pl. 5.2). Beetle larvae (weevils)
(Coleoptera: Curculionidae)   58

48 With distinct pairs of prolegs (II.14)   21
–   Without distinct pairs of prolegs, sometimes with a
single proleg-like projection on the last segment. Beetle
larvae (Coleoptera) (part)   62

49 In pods; with a spatula (II.12). Cecidomyiidae   50
–   Not in pods   51

50 Whitish yellow; non-jumping; feeding on the pod wall;
usually several in a pod; pods can become yellow,
swollen and distorted ("bladder pods"); up to 2 mm;
May–October (pl. 5.5)
*brassica pod midge *Dasineura brassicae* (Winnertz)
–   Not like this; feeding on fungi within the pod; usually
pinkish   other Cecidomyiidae

51 Last segment with 2 spiracles (breathing holes)
surrounded by a ring of 6–16 fleshy lobes; usually in or on
roots, but occasionally in stems, leaf stalks or midribs   52
–   Last segment with fewer than 6 fleshy lobes
surrounding the two spiracles; never in roots   56
(Larvae not fitting this couplet or the ones that follow on from it are
"other fly larvae".)

spiracle

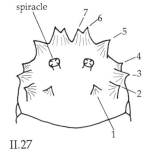

II.27

52 Last segment with a ring of 12–16 fleshy lobes around
the spiracles (view from above, II.27); last segment
divided by a ridge into an upper area with the spiracles
and a lower area with the anus; whitish; up to 10 mm.
Root flies (Anthomyiidae)   53
(Most individuals will be the pest species cabbage root fly *Delia
radicum* (L.). The following couplets can distinguish some of the less
frequent species for larvae beyond the first instar, that is, those in
which the spiracles behind the head have finger-like branches.)
–   Last segment with a ring of 6 fleshy pointed lobes
(II.13); grey or greyish brown; head retractable into the
body; feeding on roots; up to 40 mm
leatherjackets (Tipulidae)

53 Ring of lobes around the spiracles on the last abdominal
    segment with the 7th lobe from both sides fused with
    each other to form a twin-pointed lobe in the middle
    (view from above)   *Botanophila fugax* (Meigen)
    (This species used to be known as *Pegohylemyia fugax* (Meigen).)
–   Ring of lobes around the spiracles on the last abdominal
    segment with the 7th lobes separate   54

54 Ring of lobes around the spiracles on the last abdominal
    segment with the 6th and 7th lobes fused to form a
    twin-pointed lobe (II.27); spiracles behind the head with
    8–13 finger-like branches; up to 7 mm (pl. 5.6)
                       cabbage root fly Delia radicum (L.)
    (This species has been known as *Delia brassicae, Erioischia brassicae,
    Chortophila brassicae* and *Hylemyia brassicae*.)
–   Ring of lobes around the spiracles on the last abdominal
    segment with the 6th and 7th lobes separate; front
    spiracles often with fewer or more branches   55

55 6th and 7th lobes of similar size; 7th lobes easily visible
    between 6th lobes (view from above); front spiracles
    with 8–16 branches; up to 10 mm
                       turnip root fly *Delia floralis* (Fallén)
–   7th lobes small and inconspicuous; front spiracles with
    5–8 branches; up to 8 mm   *Delia florilega* (Zett.)
                       and bean seed fly *Delia platura* (Meigen)
    (These species used to be known as *Hylemya trichodactyla* (Rond.) and
    Hylemya cilicrura (Rond.) respectively. They cannot be separated
    reliably as larvae. It is best to rear them and identify the adults from
    Key IV. See p. 57 for advice on rearing.)

56 Front spiracles (breathing holes behind the head) well
    separated, on the sides of the body; whitish; living in
    blister-like externally visible mines in leaves and
    occasionally leaf stalks; up to about 4 mm
                       Scaptomyza flava (Fallén) (Drosophilidae)
    (This species used to be known as *Scaptomyza apicalis* Hardy.)
–   Front spiracles close together on the upper surface;
    whitish; living in mines in leaves, leaf stalks and stems.
    Agromyzidae   57

57 Hind spiracles (breathing holes at rear of body) each with
    6–10 pores; mines in leaves, rarely leaf stalks; up to 3.5
    mm   *Phytomyza horticola* Goureau
–   Hind spiracles each with 25–30 minute pores; mines in
    leaf veins, midribs, leaf stalks and stems; up to 6 mm
              cabbage leaf miner Phytomyza rufipes Meigen

58 In pods; feeding on seeds; usually 1 per pod; up to 5
    mm (pl. 5.2)   *cabbage seed weevil
                       Ceutorhynchus assimilis* (Paykull)
–   Not in pods   59

59   In round galls on the roots or stem bases; usually 1 larva
     per gall; up to 4 mm   turnip gall weevil
     Ceutorhynchus pleurostigma (Marsham)
  – Not in galls on the roots or stem bases; in mines in
     stems, leaf stalks and leaves   60

60   Mines at least partly in the leaf blade, may extend into
     veins and midribs; May–July; up to 3.5 mm
     turnip stem weevil *Ceutorhynchus contractus* (Marsham)
  – Mines not in the leaf blades, restricted to midribs, leaf
     stalks and stems   61

61   Abdominal segments with 3 transverse folds above and
     a single fold below; up to 4 mm
     Baris laticollis (Marsham)
  – Abdominal segments with 4 transverse folds above and
     two folds below (II.28); up to 5 mm
     Ceutorhynchus species

II.28

(These *Ceutorhynchus* species look the same. They can be identified
by rearing them to adults and using Key III. The three commonest
species may be separable by their time of appearance or type of
damage.

*Cabbage stem weevil *Ceutorhynchus quadridens* (Panzer): late
March–July, most obvious May–June; common on cabbage and rape;
weakens rape stems, but does not split stems or encourage side
branches.

*Rape winter stem weevil *Ceutorhynchus picitarsis* Gyllenhal: late
September–May, most obvious November–May; common on winter
rape, but not spring rape or cabbage; splits the stems and kills the
main shoot of rape during the winter or stunts the plant, often
encouraging side branches.

Ceutorhynchus rapae Gyllenhal: May–August, uncommon, not present
in large numbers)

62   Found mining within leaves, leaf stalks or stems   63
  – Found at the surface of roots or mining within roots   65

63   Abdomen with a pair of jointed, backwards-pointing
     prongs at the end (II.26); body, apart from head, dirty
     white with brown plates on the upper surface; up to 10
     mm; mining in stems; November–April
     turnip mud beetle *Helophorus* species (Hydrophilidae)
  – Abdomen without a pair of jointed, backward-pointing
     prongs at the end, but sometimes with unjointed hooks
     64

64   Found in blister-like leaf-blade mines up to 20 mm; last
     abdominal segment without hooks; body, apart from
     head, yellow with dark brown plates; up to 6 mm; June–
     July (pl. 5.4)   *large striped flea beetle or turnip flea
     beetle larva *Phyllotreta nemorum* (L.) (Chrysomelidae)

  – Found in mines in stems, leaf stalks and midribs; last abdominal segment with 2 upwardly curved hooks; head black; upperside plates next to the head and on the end of the abdomen greyish black; rest of body creamy white with grey plates; up to 8 mm; October–May (pl. 5.3) *cabbage stem flea beetle *Psylliodes chrysocephala* (L.) (Chrysomelidae)

65 Abdomen with a distinct pair of prongs at the end (II.26); often active predators 66
  – Abdomen without a distinct pair of prongs at the end 67

II.29

66 Legs consisting of 4 segments (II.29) +rove beetles (Staphylinidae)
  – Legs consisting of 5 segments +ground beetles (Carabidae)

67 Body, apart from head, creamy white, curled in the shape of a C when at rest; last segment of abdomen larger than the previous segments; up to 60 mm when straightened out chafers and cockchafers (Scarabaeidae)
  – Body straight when at rest; last segment of abdomen not larger than the previous segments 68

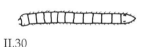

II.30

68 Skin tough and yellow to yellow-brown, without plates; up to 37 mm; abdomen pointed at the end (II.30) wireworms (Elateridae)
  – Skin soft and creamy white to yellow with dark plates; up to 6 mm. Chrysomelidae: *Phyllotreta* species 69

II.31

69 2nd and 3rd segments (counting from behind the head) with 8 small plates on the middle of the upper surface; most of body creamy white to yellow; up to 4 mm barley flea beetle *Phyllotreta vittula* Redt.
  – 2nd and 3rd segments with 6 small plates on the middle of the upper surface (II.31); most of body creamy white; up to 6 mm 70

(Larvae of *Phyllotreta* species are hard to identify. The following couplets separate full-grown larvae (4–6 mm) of some of the commoner species.)

II.32

70 Last (9th) abdominal segment with a projecting hooked lobe on the end (II.32); up to 6 mm *Phyllotreta atra* (F.) and *Phyllotreta cruciferae* (Goeze)

(These two species cannot be separated as larvae. It is necessary to rear them to adults and then use Key III in order to separate them. See p. 57 for advice on rearing.)
  – Last abdominal segment without a projecting hooked lobe 71

71 Last abdominal segment with the upperside plate whitish and smooth, without any dent in the middle; up to 6 mm   *Phyllotreta nigripes* (F.)
–   Last abdominal segment with the upperside plate brown to black and at least partly roughened or dented in the middle   72

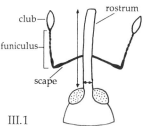

III.1

72 Last abdominal segment with the upperside plate roughened on less than half the area; up to 6 mm
          Phyllotreta consobrina (Curtis)
     and *Phyllotreta diademata* (Foudras)
(These two species cannot be separated with certainty as larvae. It is necessary to rear them to adults and identify them from Key III. See p. 57 for advice on rearing.)
–   Last abdominal segment with the upperside plate roughened on more than half the area and with a dented area in the middle; up to 5 mm
          Phyllotreta undulata Kutschera

Key III. Adult beetles (Coleoptera)

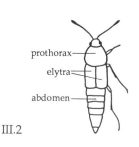

III.2

If a beetle is identified as "other beetles" or if it does not exactly fit this key and the illustrations, see Unwin (1984). Identifications to family in this key may not include less common species in that family, so it should not be assumed that beetles keying out as "other beetles" will necessarily belong to families other than those mentioned here.

1   Head longer than it is wide (view from above, exclude antennae and jaws), extending in front of the eyes in the form of a snout-like rostrum (III.1). Weevils (Curculionidae) (also see Morris, 1991)   2
–   Head not longer than it is wide   3

2   Antennae with the 1st segment (scape) longer than the next 2 segments together, usually sharply angled (III.1); rostrum at least 3 times as long in front of the eyes as it is wide just in front of the eyes (arrows in III.1 indicate which lengths to measure)   21
–   Not like this
          other weevils (Curculionidae) (see Morris, 1991)
(Several weevils from other crops, such as *Apion* and *Sitona* species, will key out here.)

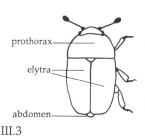

III.3

3   Elytra shorter than the abdomen, leaving at least 3 abdominal segments fully exposed (III.2); the inner edges of the elytra in contact right up to the ends and not diverging (III.2). Rove beetles (Staphylinidae)   4
–   Not like this; usually not more than the last 2 abdominal segments fully exposed (III.3)   6

III.4

III.5

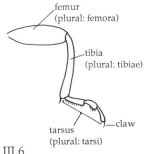

femur
(plural: femora)

tibia
(plural: tibiae)

claw

tarsus
(plural: tarsi)

III.6

short segment

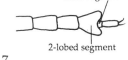

2-lobed segment

III.7

4   Body black and reddish brown or yellow; legs and antennae yellowish; body with a streamlined elongate-oval shape (III.2); up to 4 mm; runs fast over the plant; predator of aphids. *Tachyporus* species   5
–   Not like this   +other rove beetles (Staphylinidae)

5   Abdomen black, apart from reddish brown hind margins of segments; elytra reddish brown, apart from black at the front margin; head and prothorax black, apart from some yellowish mouthparts. (Identification should be checked by an expert.)
                                                    +*Tachyporus hypnorum* (F.)
–   Not like this   +other *Tachyporus* species

6   Antennae with a club at the tip, formed either from symmetrically expanded segments (III.4) or from a series of asymmetrically projecting plates (III.5)   7
–   Antennae without a club at the tip, of uniform width, or gradually widening toward the tip, or with projections along most of the length   16

7   Antennal club formed from asymmetrically projecting plates (III.5)   chafers and cockchafers (Scarabaeidae)
–   Antennal club formed from symmetrically expanded segments (III.4)   8

8   Tarsi with 3 distinct segments (III.6) (high magnification and good lighting are needed; claws are not counted and very short segments immediately following 2-lobed segments (III.7) are also not counted, since they may be hidden)   9
–   Tarsi with 4 segments   10
–   Tarsi with 5 segments   14
    (Note that there are three descriptions to choose between here, rather than the usual two.)

9   Elytra (III.3) with red/orange/yellow and black markings; body oval-shaped; less than 10 mm; often feeding on aphids   +ladybirds (Coccinellidae)
                                              (see Majerus & Kearns, 1989)
    (Usually, *Adalia bipunctata* (L.) has elytra that are red with a total of 2 black spots, *Coccinella septempunctata* L. is red with 7 black spots, and *Coccinella undecimpunctata* L. is red with 11 black spots, but there is much variation in pattern.)
–   Not like this   other beetles

10 Antennal club compact and formed from 3 segments (III.4); elytra squarish at the end, as if cut off (truncate), exposing the end of the abdomen (III.3); often in flowers. Nitidulidae   11
–   Not like this   other beetles

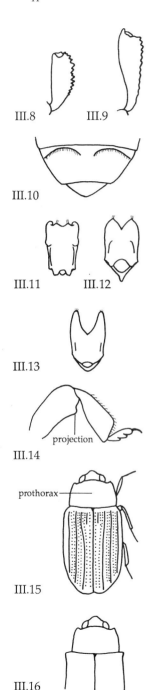

III.8   III.9

III.10

III.11   III.12

III.13

projection

III.14

prothorax

III.15

III.16

11   Body black with or without metallic blue, green or bronze, without any other markings; front tibiae (III.6) with teeth along the edge (III.8, III.9); underside of last abdominal segment with 2 semi-circular indentations (III.10) (stretch the abdomen to expose this segment fully as the indentations are often overlapped by the previous segment); 1.5–3.0 mm (pl. 1.5). Pollen beetles or blossom beetles (*Meligethes* species)   12

(Most individuals are likely to be *Meligethes aeneus*, but other species are often present. The genitalia of the males can be used to confirm species. Hook the genitalia out through the slit in the end of the abdomen. In males there will be a brown part, the tegmen, resembling those in III.11, III.12 and III.13. It is not usually possible to distinguish the sexes without pulling out the genitalia.)

–   Not like this   other Nitidulidae

12   Front tibiae with teeth tending to increase in size towards the apex (the end away from the body), but irregular in size and spacing (III.8); body black without metallic colour; 1.7–2.5 mm; legs and antennae yellowish brown; not thought to breed on brassicas; male tegmen as in III.11   *Meligethes nigrescens* Stephens

–   Front tibiae with evenly-spaced small teeth, gradually increasing in size towards the apex (III.9); body black with metallic blue, green or bronze; 1.5–2.7 mm   13

13   Lower hind edge of middle femora (= plural of femur) with an angular projection at the end away from the body (III.14); legs and antennae yellowish brown; male tegmen as in III.12   *Meligethes viridescens* (F.)

–   Lower hind edge of middle femora without a projection; legs and antennae greyish brown to black; male tegmen as in III.13 (pl. 1.5)   **Meligethes aeneus* (F.)

14   Body up to 6 mm long; front corners of prothorax projecting forwards (III.15); head much narrower than prothorax; elytra with ridges and rows of small pits, the 2nd row of pits from the centre line branching just before the front margin (III.15); brownish with yellow legs; often coated with soil particles. Turnip mud beetles (Hydrophilidae: *Helophorus* species)   15

–   Not like this   other beetles

15   Front corners of elytra angled (III.16); prothorax only slightly convex when viewed from the side (III.17); 4–6 mm   *Helophorus rufipes* (Bosc d'Antic)

–   Front corners of elytra rounded (III.15); prothorax distinctly convex when viewed from the side (III.18); 3.5–5 mm   *Helophorus porculus* Bedel

16   Hind femur (III.6) slender and less than twice as broad as the middle femur   17

–   Hind femur enlarged and at least twice as broad as the middle femur; these beetles can jump. Flea beetles (Chrysomelidae: Halticinae)   29

III.17

17 Tarsi (III.6) with 4 distinct segments (high magnification and good lighting are needed; claws are not counted and very short segments immediately following 2-lobed segments (III.7) are also not counted, since they may be hidden) 18
– Tarsi with 5 distinct segments 19

III.18

18 Body oval, metallic blue to blue-green; antennae and legs black to dark reddish brown; first antennal segment black with a dark reddish brown spot near the tip; 3–4 mm   mustard beetle or watercress beetle
      Phaedon cochleariae (F.) (Chrysomelidae)
– Not like this   other beetles

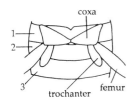

III.19

19 Hindleg with trochanter longer than the maximum width of the femur (III.19); hind coxa fused to body and extending from the front to the hind margin of the first abdominal segment (III.19)
     +ground beetles (Carabidae) (see Forsythe, 1987)
– Not like this 20

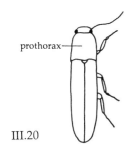

III.20

20 Prothorax with sharp hind corners pointing backwards (III.20); prosternum (underside of prothorax) projecting backwards between front legs (III.21); these beetles can jump when lying on their backs
      click beetles (Elateridae)
– Prothorax without sharp hind corners pointing backwards; prosternum not projecting backwards between front legs   other beetles

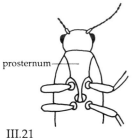

III.21

21 Rostrum more than 5 times as long in front of the eyes as it is wide just in front of the eyes (arrows in III.1 indicate which lengths to measure); prothorax (III.3) with a notch in the middle of the underside front edge where the rostrum rests; body not shiny and may have scales (scales are greyish white and like flattened hairs about 60 μm long, magnification and good lighting needed); up to 3.5 mm (weevil sizes exclude the rostrum). *Ceutorhynchus* species 22
– Rostrum up to 5 times as long in front of the eyes as it is wide just in front of the eyes; prothorax without a notch in the middle of the underside front edge; body shiny, without scales; at least 2.8 mm. *Baris* species 28

22 Antennal funiculus (III.1) with 6 segments; body greyish black; about 1.5–2.0 mm
      Ceutorhynchus floralis (Paykull)
– Antennal funiculus with 7 segments 23

23 Body 1.4–1.8 mm long, excluding rostrum and antennae; body and legs greyish black   turnip stem weevil
      Ceutorhynchus contractus (Marsham)
– Body at least 2.0 mm long 24

24   Tarsi reddish yellow all over, different in colour from the
     black femora and tibiae (III.6)   25
–   Tarsi black to dark brown, similar in colour to the
     femora and tibiae, sometimes the lobes of the 3rd tarsal
     segment (counting away from the body) light brown   26

25   Elytral interstices (smooth stripes between the rows of
     pits) with fine hairs and scattered greyish white scales
     all over, the scales often concentrated centrally near the
     front margin; 2.3–3.5 mm (pl. 1.7)   *cabbage stem
                  weevil *Ceutorhynchus quadridens* (Panzer)
–   Elytral interstices with fine hairs all over, any scales on
     them restricted to a small central patch just behind the
     front margin; 2.4–3.5 mm (pl. 1.8)   *rape winter
          stem weevil *Ceutorhynchus picitarsis* Gyllenhal

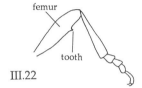

femur

tooth

III.22

26   Elytral interstices (smooth stripes between the rows of
     pits) with fine hairs all over, but without any scales
     except sometimes a few centrally near the front margin;
     hind femora with a distinct tooth underneath (III.22);
     body appearing black (viewed from above with the
     naked eye); 2.2–3.2 mm   turnip gall weevil
                  Ceutorhynchus pleurostigma (Marsham)
–   Elytral interstices with fine hairs and greyish white
     scales all over; hind femora with or without a tooth
     underneath; body appearing grey (viewed from above
     with the naked eye)   27

claw

III.23

27   Elytral interstices centrally varying from 1–2 rows of
     scales along their length; tarsal claws not toothed; hind
     femora without a tooth underneath; 2.0–3.5 mm (pl. 1.6)
          *cabbage seed weevil *Ceutorhynchus assimilis* (Paykull)
–   Elytral interstices centrally varying from 2–3 rows of
     scales along their length; tarsal claws toothed,
     sometimes giving the appearance of a 3rd central claw
     (III.23) (needs high magnification); hind femora with
     a small, sometimes indistinct, tooth underneath;
     2.0–3.5 mm   *Ceutorhynchus rapae* Gyllenhal

28   Body shiny black; 2.8–4.0 mm   *Baris laticollis* (Marsham)
–   Body shiny dark blue to green; tarsi, head below eyes
     and at least the base of the antennae reddish brown;
     3.5–6.0 mm   *Baris chlorizans* Germar

     (*B. chlorizans* is doubtfully recorded from Britain, but it is a pest of
     rape on the Continent. If you find it, have the identification checked,
     perhaps at your local museum. Your nearest ADAS office
     (Agricultural Development & Advisory Service) may be interested if
     you find this species.)

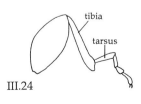

tibia

tarsus

III.24

29   Hind tarsus joins the tibia away from the tip of the tibia
     (III.24); antennae with 10 segments; body usually black
     with metallic blue-green. *Psylliodes* species   30

–   Hind tarsus joins the tibia at the tip of the tibia;
     antennae with 11 segments   31

30 Head between the antennae yellowish brown, the same
    colour as the front legs; body black with metallic blue-
    green; 3.0–5.0 mm (pl. 1.4)
        *cabbage stem flea beetle *Psylliodes chrysocephala* (L.)
–   Head between the antennae black, much darker than the
    front legs; less than 3.5 mm   other *Psylliodes* species

31 Prothorax and elytra (III.3) black with or without
    metallic blue-green, sometimes with a yellow band
    along each of the elytra; up to 3 mm; eats many small
    holes in leaves of young plants; usually April–October.
    Phyllotreta species   32
–   Not like this   other flea beetles (Chrysomelidae: Halticinae)

32 Elytra each with a yellow band (pl. 1.2, 1.3)   33
–   Elytra without yellow markings (pl. 1.1)   35

33 Tibiae all yellow; 2.4–3.0 mm (pl. 1.2)   *large striped
    flea beetle or turnip flea beetle *Phyllotreta nemorum* (L.)
–   Tibiae dark brown to black along at least half the length 34

34 2.0–3.0 mm; yellow bands not, or only slightly, indented
    by the black bumps on the front outer corners of the
    elytra (pl. 1.3)   *small striped flea beetle
                                  Phyllotreta undulata Kutschera
–   1.3–1.8 mm; yellow bands sharply indented by the black
    bumps on the front outer corners of the elytra (III.25); a
    pest of cereals, but also found on crucifers
                        barley flea beetle *Phyllotreta vittula* Redt.

III.25

35 All antennal segments black   36
–   Antennal segments 2, 3, and sometimes the tip of 1
    (counting away from the body), reddish brown, the
    others black   37

36 Pits on the prothorax deeper, their widths greater than
    the gaps between them (adjust the lighting so that the pits
    appear as black dots on a bright reflecting background);
    sometimes (males, but it is not easy to distinguish the
    sexes) with antennal segments 4 and 5 wider than 6;
    1.8–2.5 mm   *Phyllotreta consobrina* (Curtis)
–   Pits on the prothorax shallower, their widths less than or
    the same as the gaps between them when viewed as
    above; antennal segments 4 and 5 never wider than 6;
    2.0–2.6 mm   *Phyllotreta nigripes* (F.)
    (The separation of females of these two species is difficult. If you
    wish to distinguish them with certainty, consult an expert.)

37 Body black with metallic blue-green (view with the
    naked eye in daylight or background light); 1.8—2.5 mm
                               Phyllotreta cruciferae (Goeze)
–   Body shiny black without metallic her blue-green when
    viewed as above   38

38 Elytral pits of the same depth and spacing as on the
prothorax, in a confused pattern all over, never in
irregular longitudinal rows; 1.6–2.0 mm; rare
small black flea beetle *Phyllotreta aerea* Allard
– Not like this   39

39 Pits on top of the head limited to a transverse zone
between the eyes (III.26) (view from in front: the absence
of pits at the base of the head may be hidden if the head
is tilted back into the prothorax); elytral pits often
without any irregular rows; elytral pits deeper and
spaced further apart than pits on the prothorax; 1.5–2.4
mm; rare
crown flea beetle *Phyllotreta diademata* (Foudras)
– Pits all over the top of the head; elytral pits in irregular
rows, at least in places; elytral pits often similar in depth
and spacing to pits on the prothorax, but sometimes
deeper and spaced further apart; 1.7–2.6 mm (pl. 1.1)
Phyllotreta atra (F.)

(The separation of these two species is difficult. If you wish to
distinguish them with certainty, consult an expert.)

III.26

Key IV. Adult flies (Diptera)

A wide variety of flies can be found on cabbages and
rape, many of them having no particular association with
the plants. This key will identify flies particularly associated
with cabbages and rape and some common groups of flies
found around these plants. Family identifications can be
checked in the key by Unwin (1981), which should also be
used for flies identified as "other flies" and flies not exactly
fitting this key and illustrations. Identifications to family in
this key may not include less common species in that family,
so it should not be assumed that flies keying out as "other
flies" will necessarily belong to families other than those
mentioned here.

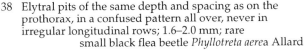

IV.1

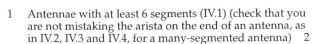

arista

IV.2

1 Antennae with at least 6 segments (IV.1) (check that you
are not mistaking the arista on the end of an antenna, as
in IV.2, IV.3 and IV.4, for a many-segmented antenna)   2
– Antennae with fewer than 6 segments (IV.2, IV.3, IV.4)   6

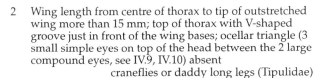

arista

IV.3

2 Wing length from centre of thorax to tip of outstretched
wing more than 15 mm; top of thorax with V-shaped
groove just in front of the wing bases; ocellar triangle (3
small simple eyes on top of the head between the 2 large
compound eyes, see IV.9, IV.10) absent
craneflies or daddy long legs (Tipulidae)
– Not like this   3

IV.4

3 Wings with 3 or 4 veins reaching the wing margin,
which is fringed with hairs (count all the veins, whether
or not they are pigmented); up to 5 mm, excluding
ovipositor (egg-laying tube). Gall midges
(Cecidomyiidae)   4

IV.5

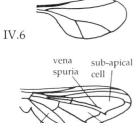

IV.6

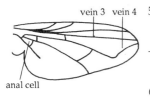

vena spuria   sub-apical cell

anal cell   discal cell

IV.7

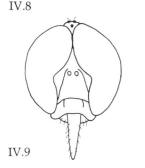

vein 3   vein 4

anal cell

IV.8

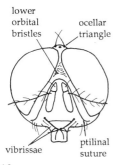

IV.9

lower orbital bristles   ocellar triangle

vibrissae   ptilinal suture

IV.10

(Gall midges are hard to identify. The following couplets separate the three commonest species associated with cabbages and rape, but others may be found, often coming from nearby crops of other plants, particularly if traps are used.)

– Not like this   other flies

4 Wing with long vein running from the base all the way to the tip (IV.5); antenna with 14 segments (note that in females, distinguished by a tubular ovipositor at the end of the abdomen, each antenna is like a string of 14 beads, but in males the last 12 segments of the antenna are narrow in the middle and so each antenna is like a string of 26 beads); abdomen yellow; hind part of thorax with 2 shiny black grooves along it; 1.5–2.0 mm, excluding ovipositor; May–October   swede midge *Contarinia nasturtii* (Kieffer)

– Wing with long vein from the base ending just before the tip (IV.6); antenna with variable number of segments from 12–16, usually 15, segments not narrow in the middle in males; 1.5–2.0 mm, excluding ovipositor   5

5 Abdomen pink to red, sometimes with brownish black markings; lays eggs on or in pods; May–August (on rape) (pl. 4.5) *brassica pod midge *Dasineura brassicae* (Winnertz)

– Abdomen yellowish brown; lays eggs in flowers brassica flower midge *Gephyraulus raphanistri* (Kieffer)

6 Wing with anal cell long and pointed, reaching at least two-thirds of the way to the wing margin (a cell is an area of the wing enclosed by veins or by veins and the wing base) (IV.7)   7

– Wing with anal cell short or absent, not reaching more than two-thirds of the way to the wing margin (IV.8)   8

7 Wing with sub-apical and discal cells closed by a cross-vein before the wing margin, with a vena spuria (a pigmented fold resembling a vein) reaching into the sub-apical cell (IV.7); abdomen often with yellow and black markings   +hoverflies (Syrphidae)
(see Stubbs & Falk, 1983; Gilbert, 1986)

– Not like this   other flies

8 Antenna with arista at the tip (IV.2) or arista absent   9

– Antenna with arista on the upper edge (IV.3, IV.4)   10

9 Eyes with a notch at the front (IV.9); body with bristles; head usually almost spherical; predator of small insects
+Empididae (see Collin, 1961)

– Not like this   other flies

10 Wing with anal cell short or absent, not reaching more than half of the way to the wing margin (IV.8); antennae in a hollow in the face with a ptilinal suture around it (IV.10)   11

– Not like this   other flies

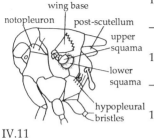

wing base

notopleuron

post-scutellum

upper squama

lower squama

hypopleural bristles

IV.11

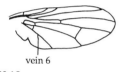

femur

tarsus

tibia

IV.12

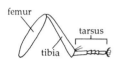

vein 6

IV.13

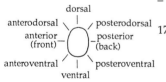

dorsal

anterodorsal   posterodorsal

anterior (front)   posterior (back)

anteroventral   posteroventral

ventral

IV.14

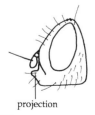

projection

IV.15

11 Wing with vein 4 bent sharply upwards, reaching the margin close to vein 3 or joining vein 3 (IV.8)   12
–   Wing with vein 4 not sharply bent, usually straight   14

12 Thorax with hypopleural bristles (just above the hind legs) (IV.11)   13
–   Thorax without hypopleural bristles   Muscidae

13 Thorax with post-scutellum strongly developed (IV.11)   +Tachinidae
–   Thorax with post-scutellum absent or weakly developed   Calliphoridae, Sarcophagidae

14 Antennae with segment 3 (counting from the base) elongate, considerably longer than wide (IV.3)   15
–   Antennae with segment 3 rounded in profile (IV.4)   31

15 Head with vibrissae present (black bristles, longer and thicker than the others, by the mouth) (IV.10)   16
–   Head without vibrissae   30

16 Hind tibia (IV.12) with bristles in the four-fifths nearest the base at least as long as the width of the tibia; lower squama (flap just behind the wing base) as long as or longer than the upper squama (IV.11); wing with vein 6 reaching the wing margin (IV.13), although not always pigmented right to the tip; body usually grey and hairy. Anthomyiidae   17
–   Not like this   28

17 Thorax upperside greenish grey, bluish grey, brownish grey or grey, apart from, sometimes, 3 or 5 length-wise dark stripes varying from brown to black according to angle of viewing and lighting; hind tibia with more than 5 bristles on the upper surface; arista (IV.3) with no hairs longer than twice the width of the base of the arista; eyes occupying at least two-thirds of the height of the head in side view; 3–8 mm   18

(Many of the individuals keying out here are likely to be cabbage root fly *Delia radicum*, but several other species are common and look the same to the naked eye. They are difficult to separate, so particular care is needed with the following couplets. The positions of bristles on the legs are critical. These positions are defined with reference to an imagined cross-section of the leg when it is extended out sideways (IV.14). For the names of the parts of the leg, see IV.12.)

–   Not like this   other Anthomyiidae

18 Lower margin of face projecting beyond the rest of the head (excluding antennae) (IV.15)   19
–   Lower margin of face not projecting as far as the rest of the head; middle tibia without an anteroventral bristle near the middle   20

19 Middle tibia with an anteroventral bristle near the
    middle; notopleuron (IV.11) with hairs between the two
    strong bristles; about 5 mm   *Paregle audacula* Harris
    (This species used to be known as *Paregle radicum* (L.).)
  – Not like this   other Anthomyiidae

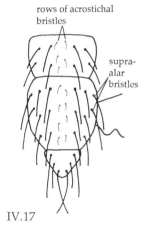

IV.16

rows of acrostichal
bristles

supra-
alar
bristles

IV.17

20 Notopleuron (IV.11) with 2 or more fine hairs between
    the 2 strong bristles; outer half of vein along the front
    margin of the wing with short black bristles underneath,
    almost to the wing tip (IV.16) (these are not easy to see,
    particularly when the vein is dark: the bristles may be
    sparse in old and worn specimens; high magnification
    and good light are needed); thorax with the front supra-
    alar bristle (above the wing base) more than half as long
    as the hind one (IV.17); 4–7 mm   21
  – Notopleuron without hairs between the 2 strong
    bristles; outer half of vein along the front margin of the
    wing without short black bristles underneath (IV.18);
    3–8 mm   22

21 Eyes nearly meeting, the gap between them at the
    narrowest part less than the width of the ocellar triangle
    (IV.10); hind femur with hairs longest and densest
    anteroventrally near the base
                 males of cabbage root fly Delia radicum (L.)
  – Eyes further apart, the gap between them at the
    narrowest part much greater than the width of the
    ocellar triangle; hind femur underside with a single fine
    erect hair at the base (pl. 3.4)
                 females of cabbage root fly Delia radicum (L.)
    (This species has also been known as *Delia brassicae, Erioischia
    brassicae, Chortophila brassicae* and *Hylemyia brassicae*.)

22 Thorax with the front supra-alar bristle (above the wing
    base) more than half as long as the hind one (IV.17);
    5–8 mm   23
  – Thorax with the front supra-alar bristle less than half as
    long as the hind one; 3–6 mm   24

IV.18

23 Eyes nearly meeting, the gap between them at the
    narrowest part less than the width of the ocellar triangle
    (IV.10)   males of turnip root fly *Delia floralis* (Fallén)
  – Eyes further apart, the gap between them at the
    narrowest part much greater than the width of the
    ocellar triangle; hind femur underside with a single fine
    erect hair at the base
            females of turnip root fly *Delia floralis* (Fallén)

24 Eyes nearly meeting, the gap between them at the
    narrowest part less than the width of the ocellar triangle
    (males) (IV.10)   25
  – Eyes further apart, the gap between them at the
    narrowest part much greater than the width of the
    ocellar triangle (females)   27

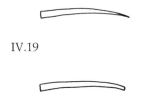

IV.19

IV.20

25 Front tibia with a sharply pointed posteroventral bristle at the tip (IV.19); thorax with one pair of strong acrostichal bristles (IV.17) near the front margin, the other acrostichals much shorter and finer hairs; 4–6 mm
                                        males of *Botanophila fugax* (Meigen)
(This species used to be known as *Pegohylemyia fugax* (Meigen).)
  – Front tibia with a blunt posteroventral bristle at the tip (IV.20)   26

26 Middle tarsus with dorsal bristles at the base longer than the width of the tarsus; 3–6 mm
                          males of bean seed fly *Delia florilega* (Zett.)
(This species used to be known as *Hylemya trichodactyla* (Rond.).)
  – Middle tarsus with dorsal bristles at the base shorter than the width of the tarsus; 3.5–5 mm
                          males of bean seed fly *Delia platura* (Meigen)
(This species used to be known as *Hylemya cilicrura* (Rond.).)

27 Hind femur with a single, fine, erect, absolutely ventral hair (near the base); arista (IV.3) with hairs in the outer half at least as long as the width of the base of the arista; 3–6 mm   females of bean seed flies
         Delia platura (Meigen) and *Delia florilega* (Zett.)
(Females of these two species cannot be reliably separated by single characters. They used to be known as *Hylemya cilicrura* (Rond.) and *Hylemya trichodactyla* (Rond.) respectively.)
  – Hind femur with at least 2 absolutely ventral hairs or bristles; arista with hairs in the outer half not as long as the width of the base of the arista; thorax with 1 pair of strong acrostichal bristles (IV.17) near the front margin, the other acrostichals much shorter and finer hairs; 4–6 mm   females of *Botanophila fugax* (Meigen)
(This species used to be known as *Pegohylemyia fugax* (Meigen).)

28 Head with fine pale hairs on the sides behind the eyes; lower squama (IV.11) not more than half the length of the upper squama, often reduced to a narrow fringed strip. Dung-flies and their relatives (Scathophagidae) 29
  – Head without fine pale hairs on the sides behind the eyes   other flies

29 Antennae black; hairs on the arista (IV.3) more than twice as long as the width of the base of the arista; upperside of hind tibia (IV.12) with at least 10 bristles; often (males) with golden-yellow hairs on the abdomen and legs; often common on cabbages and oilseed rape, breeding on dung nearby (pl. 3.5)
                +yellow dung-fly *Scathophaga stercoraria* (L.)
  – Not like this   other Scathophagidae

30 Head spherical; wing with a dark spot near the tip; wing length up to 4 mm; ant-like appearance, with narrow waist behind thorax; active with wing-waving dance
                                         Sepsis species (Sepsidae)

– Not like this   other flies

anal cell

IV.21

31 Head with lower orbital bristles curved inwards, vibrissae present (black bristles, longer and thicker than the others, by the mouth, IV.10); wing with anal cell present (small, so high magnification and good light needed), no branching of the veins in the outer half of the wing (IV.21); wing up to 3.5 mm long. Agromyzidae 32
– Not like this   33

32 Femora yellow with a few brown markings; wing 2.5–3.5 mm long (pl. 3.3)
                    *cabbage leaf miner *Phytomyza rufipes* Meigen
– Femora black with yellow knees; wing 2.2–2.6 mm long
                               Phytomyza horticola Goureau

IV.22

33 Arista hairy, with a Y-fork at the tip (IV.22); body yellow to pale brown all over; less than 3 mm (pl. 4.6)
               Scaptomyza flava (Fallén) (Drosophilidae)
    (Also referred to as *Scaptomyza apicalis* Hardy.)
– Not like this   other flies

Key V. Adult butterflies and moths (Lepidoptera)

Wing length is the length from the centre of the thorax to the furthest point of the outspread forewing, including any fringe. The key may not work for exceptionally large or small individuals, which sometimes occur.
       Identifications should be checked in standard works for butterflies and larger moths, e.g. Skinner (1984) and Higgins & Hargreaves (1983). References are given within the key for micro-moths.

1 Antennae with a solid club at the end; flying readily if disturbed during the day. Butterflies   2
– Antennae tapered, sometimes slightly plumed like a feather; usually inactive by day. Moths   4

2 Wing length more than 29 mm; upperside white with black tip and two black spots (in females) or no black spots (in males)   *large white
                   or cabbage white *Pieris brassicae* (L.) (Pieridae)
– Wing length less than 27 mm; upperside white with black markings   3

3 Hindwing underside pale yellow with veins outlined in grey-green; forewing upperside white with grey-black markings along the veins and 2 black spots (in females) or 1 or no black spots (in males)
                   green-veined white *Artogeia napi* (L.) (Pieridae)
    (This species is also widely known as *Pieris napi* (L.).)

- Hindwing underside pale yellow with a dusting of greyish scales with veins not outlined; forewing upperside white with black tip and 2 black spots (in females) or 1 or no black spots (in males)
    *small white *Artogeia rapae* (L.) (Pieridae)
    (This species is also widely known as *Pieris rapae* (L.).)

4   Width of thorax, including hair, less than 3.5 mm; wing length up to 17 mm   5
- Width of thorax, including hair, more than 4 mm; wing length at least 16 mm. Noctuidae   8

5   Wing length less than 11 mm   6
- Wing length more than 13 mm   7

6   Forewings with 3 pale triangular marks along the hind margin that give a diamond pattern when the wings are closed, background colour greyish; wing length about 9 mm   *diamond-back moth *Plutella xylostella* (L.) (Yponomeutidae)(see Holloway and others, 1987)
    (This species used to be known as *Plutella maculipennis* (Curt.).)
- Forewings without triangular marks, background colour very variable from mottled light brown all over to blackish brown with black specks; wing length 8–9 mm; June–August   flax tortrix moth *Cnephasia interjectana* (Haworth) (Tortricidae) (see Bradley and others, 1973)

7   Forewing pale brownish yellow with 2 brown dots in the middle and several brown lines, the strongest running from the wing tip to the middle of the hind edge; wing length 15–17 mm; May–June and August–September
    *garden pebble *Evergestis forficalis* (L.) (Pyralidae) (see Goater, 1986)
- Forewing greyish white with irregular grey lines running from the front to the hind margin and large brownish black markings; wing length 13–16 mm; April–October   garden carpet *Xanthorhoe fluctuata* (L.) (Geometridae)

8   Active and flying by day; forewing greyish brown with a white Y-shaped mark in the middle; wing length 16–26 mm   silver Y moth *Autographa gamma* (L.)
- Inactive by day, unless disturbed; forewing without a white Y-shaped mark   9

9   Eyes hairy (high magnification needed); hindwing grey to light brown, becoming lighter towards the base   12
- Eyes smooth. Adult cutworms   10

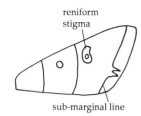

reniform stigma

sub-marginal line

V.1

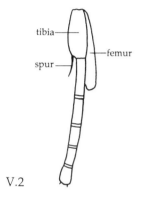

tibia

femur

spur

V.2

10 Forewing reddish brown; reniform stigma (a distinct blotch just beyond the middle of the forewing, towards the front margin, V.1) light brown without white markings; sub-marginal line (a line just inside and parallel to the outer margin of the forewing, V.1) white with a W-shaped zigzag in the middle; wing length 17–22 mm; usually May–July   tomato moth or bright-line brown-eye *Lacanobia oleracea* (L.)
– Forewing greyish brown to black; reniform stigma with white markings; wing length 19–25 mm   11

11 Front leg with a brown, curved spur at the tip of the tibia (V.2); reniform stigma grey, edged with white lines; forewing greyish brown; wing length 19—25 mm; May–October   *cabbage moth *Mamestra brassicae* (L.)
– Front leg without a spur at the tip of the tibia; reniform stigma solid white, apart from a reddish grey comma-shaped mark in the centre; forewing dark greyish brown to black, sometimes with a purple tinge; wing length 19–25 mm; July–August   dot moth *Melanchra persicariae* (L.)

12 Hindwing orange with black markings; forewing light brown to grey or dark brown; wing length 25—30 mm; July–September
        large yellow underwing *Noctua pronuba* (L.)
– Hindwing white, cream or light brown, sometimes with grey to black markings; forewing light brown to grey-brown   13

13 Forewing with 1–3 solid black dashes or wedge-shaped markings running parallel to the hind margin   14
– Forewing without such markings, any black dashes or wedge-shaped markings are outlined and not solid   15

14 Forewing with 1 solid black dash (2–3 mm long) just inside the inner half of the wing; wing length 17–22 mm; May–July   heart and dart moth *Agrotis exclamationis* (L.)
– Forewing with 1–3 solid black wedge-shaped marks in the outer third of the wing, an irregular light brown band (2–4 mm wide) running across the marks from the front to hind margin of the wing; wing length 20–28 mm; March–November   dark sword-grass or dark dart *Agrotis ipsilon* (Hufnagel)

15 Hindwing with a brown border at least 2 mm wide in places, rest of wing pearly grey with brown markings along the veins; wing length 22–28 mm; January–December   pearly underwing *Peridroma saucia* (Hübner)
– Hindwing with a brown border less than 2 mm wide, rest of wing white to grey with brown markings along the veins; wing length 16–22 mm; May–June and August–September
        turnip moth *Agrotis segetum* (D. & S.)

6 Techniques

Preserving insects for identification

Fig. 17. A pinned insect.

Fig. 18. A staged insect.

Most adult butterflies and moths can be identified alive. Some insects can be identified while temporarily kept still, after being chilled in a refrigerator or anaesthetised with carbon dioxide from a gas-operated wine-bottle opener. However, in order to identify an insect with certainty and allow its identity to be checked later by others, it will usually be necessary to kill and preserve it. Adult insects can be killed in a small glass jar sealed with a cork or screw-top lid. Some killing fluid (ethyl acetate) is first poured onto blotting paper or, better still, a 1 cm layer of plaster of Paris set in the bottom of the jar. Note that ethyl acetate vapour attacks many plastics. Watkins and Doncaster (address p. 60) supply ready-made killing jars, ethyl acetate, and all the entomological equipment you are likely to need. Some chemists will supply ethyl acetate. The best plaster of Paris is that used by dentists and sold by dental supply companies (see the Yellow Pages of your telephone directory). The jar should be kept out of the sun to avoid condensation. Specimens will be anaesthetised within seconds and most insects will be dead after about half an hour, but large bees or beetles should be left for at least an hour. Large specimens should then be pinned with an entomological pin (other pins corrode) (fig. 17). Small insects that would be crushed by a large pin should be pinned with a fine headless entomological pin, which is then "staged" on a small strip of expanded polyethylene foam (from Polyformes Ltd, address p. 60) (fig. 18). A paper or card label with the place and date of capture should be pinned below every specimen. If the insect was found on a plant, the plant name should be given, and the name of the insect should be added when known. Labelling is tedious, but it is vital if your specimens are to have any scientific value. Hard-bodied adults dry out naturally and need no treatment in order to be preserved. Pinned specimens can be stored in a box floored with cork, expanded polystyrene ceiling tiles or expanded polyethylene foam (from Polyformes Ltd, address p. 60) and with a tight-fitting lid to keep out destructive mites and insects. These can also be deterred by a few crystals of paradichlorbenzene (available from chemists as a moth repellent) well sealed into a twist of muslin or paper very firmly pinned inside the box. Collections will last longer in specially constructed entomological storeboxes, but these are expensive.

Some small insects, such as thrips, can be identified only if they are mounted on microscope slides. The preparation of permanent slides of thrips is described by Mound and others (1976). Temporary mounts lasting months to years can be made with polyvinyl lactophenol (toxic). Thrips are killed in 70% alcohol (industrial methylated spirit diluted with water) and left there for at

least 24 hours to help clear them so that they can be seen
clearly under a compound microscope. Industrial
methylated spirits can only be purchased with an Excise
licence; local schools, universities or museums may have
some. Alternatively, thrips can be killed in boiling water and
mounted immediately, but it will then take longer for the
mountant to clear them. A drop of polyvinyl lactophenol is
placed on a round coverslip (diameter 13 mm and thickness
code 0) and a thrips is placed in the middle with its
underside facing upwards. A microscope slide is then
lowered onto the coverslip. When the slide is turned over so
that the coverslip is on top, the specimen will have its upper
side facing upwards. A data label should be stuck on the
slide. The mountant clears the specimen further so that after
two or three days it can be studied easily under a compound
microscope. Extra mountant should be added at the edge of
the coverslip if air bubbles appear. The mountant, coverslips
and microscope slides are available from MERCK Ltd.
(address p. 60).

Soft-bodied larvae can be killed and preserved
adequately in 70% alcohol (see above) and kept in well-
sealed glass or plastic tubes. Surgical spirit (sold by
chemists) and methylated spirits as sold by hardware shops
can be used if 70% alcohol is not available, but they are not
ideal. They are about 90–95% alcohol, which may cause
distortion of the specimen. If they are diluted with water,
they precipitate a white substance which coats the specimen.
Surgical spirit leaves an oily coating on specimens and
methylated spirits contains a dye. All three substances are
flammable. A paper data label, written in pencil, should be
dropped into the alcohol with the specimen.

Detailed information on collecting and preserving
insects is given by Oldroyd (1958) and Smithers (1982).
Many books on insects, such as Chinery (1976), contain an
introductory section on preservation of specimens.

Rearing insects

Rearing insects is one of the best ways to learn about
them. Sometimes it is necessary to rear immature insects
through to the adult stage in order to identify them. The
difficult part of rearing is usually keeping insects over
winter and persuading the adults to mate and lay eggs.
Comprehensive advice for butterflies and moths is given by
Friedrich (1986), and for most types of insect by Smithers
(1982).

Insects that feed on leaves, such as caterpillars of
butterflies, moths and sawflies, do best in well-ventilated
containers out of direct sunlight. A jam jar or sandwich box
covered with a piece of netting secured with an elastic band
works well. The insects must be regularly supplied with
fresh food. Cabbage bought from a shop may contain
residues of insecticides; organically grown cabbage may be a
safer food. Droppings and any dead caterpillars should be
removed and any moisture on the sides of the container

wiped away to prevent mould and disease. The containers should not be much bigger than the amount of food supplied or else the caterpillars will wander away and lose their food. Butterfly caterpillars from cabbages and oilseed rape will pupate on the sides of a container, but moths and sawflies will pupate in the soil, so a layer of a few centimetres of slightly moist fine soil should be provided when the larvae are fully grown. Some moths spend several weeks in the soil before pupating, so wait a while before looking for the pupae. Pupae should be kept in a container with some twigs for emerged adults to climb up. Overwintering pupae need to be kept cool, so store them in a shed or garage.

Larvae that mine within stems, leaves or roots should be kept in a container with the original plant part they were found on and a layer of fine soil for pupation. Stems and leaves will last longer if placed in a jar of water with cotton wool at the neck to prevent larvae from drowning. Choose mature larvae as they will pupate sooner. Midge larvae from flowers, pods or stems should be kept in the same way. Adults may emerge from pupae within a few weeks or they may wait until the following spring or summer. The pod midge (*Dasineura brassicae*) can spend more than two years as a pupa! Patience is needed for rearing insects.

Parasites that emerge from caterpillars and spin cocoons should be kept in small ventilated containers out of the sun. The containers should be well sealed to prevent the parasites from escaping when they emerge.

Some insects can be kept in continuous culture: butterflies and moths (Friedrich, 1986); root flies (Harris & Svec, 1966); and pollen beetles (Bromand, 1983).

Trapping insects

With very little effort, using water traps, one can catch many different types of flying insect. These can be used to monitor population sizes or to investigate responses to colours and scents. A dish, such as a washing-up bowl or an empty ice-cream container, is filled to about 2 cm below the rim with tap water containing a few drops of washing-up detergent. It is important to use an unscented detergent (such as Teepol) if you are investigating factors affecting trap catches. Insects that land on the water sink to the bottom and drown. The catch must be sieved out within a few days or it will go mouldy. The insects can be dried and pinned, or stored in 70% alcohol. Yellow traps catch many foliage-feeding insects, but other colours, such as white or blue, may catch other insects (Kirk, 1984). Colours can be varied by painting the dishes inside with gloss paint. Scent can be released from small, glass tubes attached to the side of the trap. The release rate can be controlled with a wick made from a cotton wool dental roll (from dental suppliers) (Kirk, 1985). 2-propenyl isothiocyanate is available from the Aldrich Chemical Co. Ltd and MERCK Ltd (addresses p. 60).

It is also referred to in catalogues as allyl isothiocyanate and
3-iso-thiocyanatoprop-1-ene. But beware, it is highly toxic
and a suspect cancer agent. It requires careful handling,
including use of a fume cupboard. Finch & Skinner (1982)
list several other isothiocyanates that increase trap catches,
some of which are safer. Details of the hazards from
individual chemicals are available from suppliers. Sinigrin,
which is not volatile, can be purchased for other
experiments from the Aldrich Chemical Co. Ltd; it is very
expensive. Plant extracts can be made for scent experiments
by crushing plant parts in methanol (methyl alcohol) as a
solvent (from the Aldrich Chemical Co. Ltd). Methanol is
flammable and highly toxic. Handle it in a fume cupboard
or a well-ventilated space. Other solvents, such as industrial
methylated spirits, could be tried, but the scent of some of
the ingredients might affect the insects. Crushed seeds or
leaves could be tried in open tubes without a solvent, but
the release rate would not be uniform.

If you wish to investigate the effects of colour and
scent, study papers describing such experiments for
examples of experimental design and analysis (Free &
Williams, 1978; Kirk, 1984, 1985).

Lane (1984) describes several other sampling
methods that are used to monitor insects on oilseed rape.

Writing papers

Writing up is an important part of a research project,
particularly when the findings are to be communicated to
other people. A really thorough, critical investigation that
has established new information of general interest may be
worth publishing if the insects and plants on which it is
based can be identified with certainty. Journals that publish
short papers on insect biology include the *Entomologist's
Monthly Magazine*, the *Bulletin of the Amateur Entomologists'
Society*, *The Entomologist* and, for projects with an
educational slant, the *Journal of Biological Education*. Those
unfamiliar with publishing conventions are advised to
examine current numbers of these journals to see what sort
of thing they publish, and then to write a paper along
similar lines, keeping it short, but presenting enough
information to establish the conclusions. It is then time to
consult an appropriate expert who can give advice on
whether and in what form the material might be published.
It is an unbreakable convention of scientific publication that
results are reported with scrupulous honesty. Hence it is
essential to keep detailed and accurate records throughout
the investigation, and to distinguish between certainty and
probability, and between deduction and speculation. It will
usually be necessary to apply statistical techniques to test
the significance of the findings. A book such as Fowler &
Cohen (1990) will help, but this is an area where expert
advice can contribute much to the planning, as well as the
analysis, of the work.

Some useful addresses

Suppliers of equipment and chemicals

Entomological:

Polyformes Ltd, Cherrycourt Way, Stanbridge Road,
  Leighton Buzzard, Bedfordshire LU7 8UH (for expanded polyethylene foam only)
Watkins and Doncaster, Conghurst Lane, Four Throws, Hawkhurst, Kent TN18 5ED
  (for all entomological supplies)

Microscopical and chemical:

MERCK Ltd, Hunter Boulevard, Magna Park, Lutterworth, Leicestershire LE17 4XN

Chemical:

Aldrich Chemical Co. Ltd, The Old Brickyard, New Road, Gillingham, Dorset SP8 4JL

Suppliers of biological books

New:

Field Studies Council Publications, Preston Montford, Montford Bridge,
  Shrewsbury SY4 1HW (for Unwin's keys)
Natural History Museum Publications, Cromwell Road, London SW7 5BD (for
  Handbooks for the identification of British insects)
Richmond Publishing Co. Ltd., P.O. Box 963, Slough SL2 3RS (for Naturalists'
  Handbooks and Unwin's keys)

New and secondhand:

E.W. Classey Ltd, P.O. Box 93, Faringdon, Oxfordshire SN7 7DR

Societies

Some readers may wish to join a society in order to find out more about insects
and ecology and meet people with similar interests. The names and addresses of some
established societies are given below.

The Amateur Entomologists' Society produces *A Directory for Entomologists*,
which contains a comprehensive list of names and addresses of local and national
entomological societies and recording schemes, along with many other useful
addresses.

Amateur Entomologists' Society,
  355 Hounslow Road, Hanworth, Feltham, Middlesex TW13 5JH
Association of Applied Biologists,
  AFRC Institute of Horticultural Research, Wellesbourne, Warwick CV35 9EH
British Beekeepers' Association,
  National Beekeeping Centre, National Agricultural Centre,
  Stoneleigh, Kenilworth CV8 2LZ
British Ecological Society,
  Burlington House, Piccadilly, London W1V 0LQ
British Entomological and Natural History Society,
  c/o The Royal Entomological Society of London,
  41 Queen's Gate, London SW7 5HU
International Bee Research Association, 18 North Road, Cardiff CF1 3DY
Royal Entomological Society of London, 41 Queen's Gate, London SW7 5HU

References and further reading

Finding books

Many of the books listed here will be unavailable in local and school libraries. It is possible to make arrangements to see or borrow them by seeking permission to visit the library of a local university, or by asking your local public library to borrow the work (or a photocopy of it) for you from the British Library Document Supply Centre. This may take several weeks, and it is important to present your librarian with a reference which is correct in every detail. References are acceptable in the form and order given here, namely the author's name and date of publication, followed by (for a book) the title and publisher or (for a journal article) the title of the article, the journal title, the volume number, and the first and last pages of the article.

The *Handbooks for the identification of British insects* are published by the Royal Entomological Society of London and can be bought at the Natural History Museum or by post from Natural History Museum Publications, Cromwell Road, London SW7 5BD. Reprints of the keys by Unwin are available by post from Field Studies Council Publications, Preston Montford, Montford Bridge, Shrewsbury SY4 1HW or from The Richmond Publishing Co. Ltd., P.O. Box 963, Slough SL2 3RS. Asterisks mark publications available from The Richmond Publishing Co. Ltd.

References

Anon. (1988). *Common Catalogue of Varieties of Vegetable Species.* Annex to Official Journal of the European Communities, 31 (88/C 335 A).

Anon. (1989). *Common Catalogue of Varieties of Agricultural Plant Species.* Annex to Official Journal of the European Communities, 32 (89/C 326 A).

Begon, M., Harper, J.L. & Townsend, C.R. (1990). *Ecology. Individuals, Populations and Communities* (2nd edn). Oxford: Blackwell Scientific Publications.

Benson, R.B. (1952–1958). *Handbooks for the Identification of British Insects. Hymenoptera, Symphyta.* Vol. I, parts 2(b),(c). Royal Entomological Society of London.

Blau, P.A., Feeny, P., Contardo, L. & Robson, D.S. (1978). Allyl glucosinolate and herbivorous caterpillars: a contrast in toxicity and tolerance. *Science* **200**, 1296–1298.

Blight, M.M., Pickett, J.A., Wadhams, L.J. & Woodcock, C.M. (1989). Antennal responses of *Ceutorhynchus assimilis* and *Psylliodes chrysocephala* to volatiles from oilseed rape. *Aspects of Applied Biology* **23**, 329–334.

Bradley, J.D., Tremewan, W.G. & Smith, A. (1973). *British Tortricoid Moths. Cochylidae and Tortricidae: Tortricinae.* London: The Ray Society.

Bromand, B. (1983). Possibility of continuous rearing of *Meligethes aeneus* Fabr. (Col.). *Zeitschrift für angewandte Entomologie* **96**, 419–422.

Calder, A. (1986). *Oilseed Rape and Bees.* Hebden Bridge, West Yorkshire: Northern Bee Books.

Chinery, M. (1976). *A Field Guide to the Insects of Britain and Europe* (2nd edn). London: Collins.

Chinery, M. (1986). *Collins Guide to the Insects of Britain and Western Europe.* London: Collins.

Clapham, A.R., Tutin, T.G. & Moore, D.M. (1987). *Flora of the British Isles* (3rd edn). Cambridge: Cambridge University Press.

Coaker, T.H. & Finch, S. (1971). The cabbage root fly, *Erioischia brassicae* (Bouché). Report of the National Vegetable Research Station for 1970, 23–42.

Collin, J.E. (1961). *British Flies. Vol. VI. Empididae.* Cambridge: Cambridge University Press.

Cromartie, W.J. (1975). The effect of stand size and vegetational background on the colonization of cruciferous plants by herbivorous insects. *Journal of Applied Ecology* **12**, 517–533.

Eisikowitch, D. (1981). Some aspects of pollination of oil-seed rape (*Brassica napus* L.). *Journal of Agricultural Science, Cambridge* **96**, 321–326.

Erickson, J.M. & Feeny, P. (1974). Sinigrin: a chemical barrier to larvae of the black swallowtail butterfly, *Papilio polyxenes*. *Ecology* **55**, 103–111.

Feltwell, J. (1982). *Large White Butterfly. The Biology, Biochemistry and Physiology of Pieris brassicae (Linnaeus).* The Hague: Junk.

Fenwick, G.R., Heaney, R.K. & Mullin, W.J. (1983). Glucosinolates and their breakdown products in food and food plants. *CRC Critical Reviews in Food Science and Nutrition* **18**, 123–201.

Finch, S. & Jones, T.H. (1989). An analysis of the deterrent effect of aphids on cabbage root fly (*Delia radicum*) egg-laying. *Ecological Entomology* **14**, 387–391.

Finch, S. & Skinner, G. (1982). Trapping cabbage root flies in traps baited with plant extracts and with natural and synthetic isothiocyanates. *Entomologia Experimentalis et Applicata* **31**, 133–139.

*Forsythe, T.G. (1987). *Common Ground Beetles.* Naturalists' Handbooks 8. Slough: The Richmond Publishing Co. Ltd.

Fowler, J. & Cohen, L. (1990). *Practical Statistics for Field Biology.* Milton Keynes: Open University Press.

Free, J.B. & Ferguson, A.W. (1980). Foraging of bees on oil-seed rape (*Brassica napus* L.) in relation to the stage of flowering of the crop and pest control. *Journal of Agricultural Science, Cambridge* **94**, 151–154.

Free, J.B., Ferguson, A.W. & Winfield, S. (1983). Effect of various levels of infestation by the seed weevil (*Ceutorhynchus assimilis* Payk.) on the seed yield of oil-seed rape (*Brassica napus* L.). *Journal of Agricultural Science, Cambridge* **101**, 589–596.

Free, J.B. & Williams, I.H. (1978). The responses of the pollen beetle, *Meligethes aeneus*, and the seed weevil, *Ceutorhynchus assimilis*, to oil-seed rape, *Brassica napus*, and other plants. *Journal of Applied Ecology* **15**, 761–774.

Friedrich, E. (1986). *Breeding Butterflies and Moths – a Practical Handbook for British and European Species.* Colchester: Harley Books.

Fritzche, R. (1957). *Zur Biologie und Ökologie der Rapsschädlinge aus der Gattung Meligethes.* [In German, with English summary.] *Zeitschrift für angewandte Entomologie* **40**, 222–280.

*Gilbert, F.S. (1986). *Hoverflies.* Naturalists' Handbooks 5. Cambridge: Cambridge University Press.

Gill, N.T. & Vear, K.C. (1980). *Agricultural Botany. Vol. 1. Dicotyledonous Crops* (3rd edn). London: Duckworth.

Glen, D.M., Jones, H. & Fieldsend, J.K. (1990). Damage to oilseed rape seedlings by the field slug *Deroceras reticulatum* in relation to glucosinolate concentration. *Annals of Applied Biology* **117**, 197–207.

Goater, B. (1986). *British Pyralid Moths. A Guide to their Identification.* Colchester: Harley Books.

Harris, C.R. & Svec, H.J. (1966). Mass-rearing of the cabbage maggot under controlled environmental conditions. *Journal of Economic Entomology* **59**, 569–573.

Higgins, L.G. & Hargreaves, B. (1983). *The Butterflies of Britain and Europe.* London: Collins.

Holloway, J.D., Bradley, J.D. & Carter, D.J. (1987). *CIE Guides to Insects of Importance to Man. 1. Lepidoptera.* Wallingford: CAB International.

Jones, T.H., Cole, R.A. & Finch, S. (1988). A cabbage root fly deterrent in the frass of garden pebble moth caterpillars. *Entomologia Experimentalis et Applicata* **49**, 277–282.

Jones, R.E. (1977). Movement patterns and egg distribution in cabbage butterflies. *Journal of Animal Ecology* **46**, 195–212.

Kareiva, P. (1982). Exclusion experiments and the competitive release of insects feeding on collards. *Ecology* 63, 696–704.

Kirk, W.D.J. (1984). Ecologically selective coloured traps. *Ecological Entomology* 9, 35–41.

Kirk, W.D.J. (1985). Effect of some floral scents on host finding by thrips (Insecta: Thysanoptera). *Journal of Chemical Ecology* 11, 35–43.

Kozlowski, M.W., Lux, S. & Dmoch, J. (1983). Oviposition behaviour and pod marking in the cabbage seed weevil, *Ceutorhynchus assimilis*. *Entomologia Experimentalis et Applicata* 34, 277–282.

Lane, A. (1984). *The Oilseed Rape Handbook*. Milton Keynes: The Open University Press.

Lawton, J.H. (1982). Vacant niches and unsaturated communities: a comparison of bracken herbivores at sites on two continents. *Journal of Animal Ecology* 51, 573–595.

Littlewood, S.C. (1988). Notes on the overwintering *Apanteles–Tetrastichus–Lysibia* complex (Hymenoptera: Parasitica), parasitoids of the large white butterfly. *Journal of Natural History* 22, 883–895.

*Majerus, M. & Kearns, P. (1989). *Ladybirds*. Naturalists' Handbooks 10. Slough: The Richmond Publishing Co. Ltd.

*Morris, M.G. (1991). *Weevils*. Naturalists' Handbooks 16. Slough: The Richmond Publishing Co. Ltd.

Mound, L.A., Morison, G.D., Pitkin, B.R. & Palmer, J.M. (1976). *Handbooks for the Identification of British Insects. Thysanoptera*. Vol. I, part 11. Royal Entomological Society of London.

Oldroyd, H. (1958). *Collecting, Preserving, and Studying Insects*. London: Hutchinson. (Reprinted 1970.)

Osborne, P. (1960). Observations on the natural enemies of *Meligethes aeneus* (F.) and *M. viridescens* (F.) [Coleoptera: Nitidulidae]. *Parasitology* 50, 91–110.

Osborne, P. (1965). Morphology of the immature stages of *Meligethes aeneus* (F.) and *M. viridescens* (F.) (Coleoptera, Nitidulidae). *Bulletin of Entomological Research* 54, 747–759.

Philbrick, H. & Gregg, R.B. (1967). *Companion Plants and How to Use Them*. London: Watkins.

Pickett, J.A. (1989). Semiochemicals for aphid control. *Journal of Biological Education* 23, 180–186.

Price, P.W. (1984). *Insect Ecology* (2nd edn). New York: Wiley-Interscience.

*Prŷs-Jones, O.E. & Corbet, S.A. (1991). *Bumblebees* (revised edn). Naturalists' Handbooks 6. Slough: The Richmond Publishing Co. Ltd.

Quinlan, J. & Gould, I.D. (1981). *Handbooks for the Identification of British Insects. Hymenoptera, Symphyta*. Vol. VI, part 2(a). Royal Entomological Society of London.

Richards, O.W. (1940). The biology of the small white butterfly (*Pieris rapae*) with special reference to the factors controlling its abundance. *Journal of Animal Ecology* 9, 243–288.

Rieth, J.P. & Levin, M.D. (1989). Repellency of two phenylacetate-ester pyrethroids to the honeybee. *Journal of Apicultural Research* 28, 175–179.

Root, R.B. (1973). Organization of a plant–arthropod association in simple and diverse habitats: the fauna of collards (*Brassica oleracea*). *Ecological Monographs* 43, 95–124.

*Rotheray, G.E. (1989). *Aphid Predators. Insects that Eat Greenfly*. Naturalists' Handbooks 11. Slough: The Richmond Publishing Co. Ltd.

Rothschild, M. & Schoonhoven, L.M. (1977). Assessment of egg load by *Pieris brassicae* (Lepidoptera: Pieridae). *Nature, London* 266, 352–355.

Ryan, J., Ryan, M.F. & McNaeidhe, F. (1980). The effect of interrow cover on populations of the cabbage root fly *Delia brassicae* (Wiedemann). *Journal of Applied Ecology* 17, 31–40.

Scarisbrick, D.H. & Daniels, R.W. (1986). *Oilseed Rape*. London: Collins Professional & Technical.

Simmonds, N.W. (1976). *The Evolution of Crop Plants*. London: Longman.

Skinner, B, (1984). *Colour Identification Guide to Moths of the British Isles (Macrolepidoptera)*. New York: Viking.

Skinner, G. & Finch, S. (1986). Reduction of cabbage root fly (*Delia radicum*) damage by protective discs. *Annals of Applied Biology* **108**, 1–10.

Smithers, C. (1982). *Handbook of Insect Collecting – Collection, Preparation, Preservation and Storage*. Newton Abbot: David & Charles.

Southwood, T.R.E. & Leston, D. (1959). *Land and Water Bugs of the British Isles*. London: Warne.

Strong, D.R., Lawton, J.H. & Southwood, T.R.E. (1984). *Insects on Plants. Community Patterns and Mechanisms*. Oxford: Blackwell Scientific Publications.

Stubbs, A.E. & Falk, S.J. (1983). *British Hoverflies. An Illustrated Identification Guide*. London: The British Entomological & Natural History Society.

*Unwin, D.M. (1981). A key to the families of British Diptera. *Field Studies* **5**, 513–553.

*Unwin, D.M. (1984). A key to the families of British Coleoptera. *Field Studies* **6**, 149–197.

Vaughan, J.G., MacLeod, A.J. & Jones, B.M.G. (1976). *The Biology and Chemistry of the Cruciferae*. London: Academic Press.

Ward, J.T., Basford, W.D., Hawkins, J.H, & Holliday, J.M. (1985). *Oilseed Rape*. Ipswich: Farming Press Ltd.

Williams, I.H. (1978). *Pests and Pollination of Oilseed Rape Crops in England*. Ilford, Essex: Central Association of Beekeepers.

Williams, I.H. (1980). Oil-seed rape and beekeeping, particularly in Britain. *Bee World* **61**, 141–153.

Williams, I.H. (1984). The concentration of air-borne rape pollen over a crop of oilseed rape (*Brassica napus* L.). *Journal of Agricultural Science, Cambridge* **103**, 353–357.

Williams, I.H. (1989). Pest incidence on single low and double low oilseed rape cultivars. *Aspects of Applied Biology* **23**, 277–286.

Williams, I.H. & Free, J.B. (1978). The feeding and mating behaviour of pollen beetles (*Meligethes aeneus* Fab.) and seed weevils (*Ceutorhynchus assimilis* Payk.) on oil-seed rape (*Brassica napus* L.). *Journal of Agricultural Science, Cambridge* **91**, 453–459.

Williams, I.H. & Free, J.B. (1979). Compensation of oil-seed rape (*Brassica napus* L.) plants after damage to their buds and pods. *Journal of Agricultural Science, Cambridge* **92**, 53–59.

Williams, I.H., Martin, A.P. & White, R.P. (1986). The pollination requirements of oil-seed rape (*Brassica napus* L.). *Journal of Agricultural Science, Cambridge* **106**, 27–30.

Wishart, G., Colhoun, E.H. & Monteith, E. (1957). Parasites of *Hylemya* spp. (Diptera: Anthomyiidae) that attack cruciferous crops in Europe. *Canadian Entomologist* **89**, 510–517.

*Yeo, P.F. & Corbet, S.A. (1983). *Solitary Wasps*. Naturalists' Handbooks 3. Cambridge: Cambridge University Press.

Zohren, E. (1968). Laboruntersuchungen zu Massenanzucht, Lebensweise, Eiablage und Eiablageverhalten der Kohlfliege, *Chortophila brassicae* Bouché (Diptera, Anthomyiidae). [In German, with English summary.] *Zeitschrift für angewandte Entomologie* **62**, 139–188.

Index